CONCISE PRACTICAL SURVEYING

The techniques of surveying are needed not only by surveyors themselves, but by architects, builders, engineers and estate agents. It was with this thought in mind that the authors made this book both *concise* and *practical*. The second edition has kept firmly to this successful formula. All measurements are now given in SI metric units, and descriptions of methods and equipment have been modified in the light of modern practice. The opportunity has also been taken to redraw the illustrations so as to ensure that the book maintains its reputation for providing clear, practical advice.

For those who have examinations to consider, this book fully covers the requirements of the professional bodies concerned with the subject – as well as T.E.C. Standard Units 2 and 3, the Higher National Diplomas in Building and Engineering, and the General Certificate of Education in Surveying, it is suitable for City and Guilds courses which include the elements of surveying.

The specimen field book printed on pp. 173–7 refers to the survey on the outside of the cover. Readers are invited to plot the survey from the field book and to compare it with that on the cover.

495

CONCISE PRACTICAL SURVEYING

W. G. CURTIN
F.I.C.E., F.I.Struct.E., M.Cons.E., F.I.Arb.E.
Consulting Engineer

R. F. LANE
B.Sc.(Est.Man.), F.R.I.C.S.
Assistant Director, Polytechnic of the South Bank, London
Former Head of the Department of Surveying Brixton School of Building

Edward Arnold
A division of Hodder & Stoughton
LONDON NEW YORK MELBOURNE AUCKLAND

First published in Great Britain 1955
Reprinted 1960, 1961, 1964, 1965, 1966, 1968 (with revisions)
Second edition 1970
Reprinted 1972, 1975, 1976, 1978, 1979, 1980, 1981, 1985, 1989

ISBN 0 340 04580 9

Printed and bound in Great Britain for Edward Arnold, the educational, academic and medical publishing division of Hodder and Stoughton Limited, 41 Bedford Square, London WC1B 3DQ by J. W. Arrowsmith Ltd, Bristol

FOREWORD

THIS book on surveying has been admirably conceived, and the authors—both of whom are well known to the writer—have had, not only practical experience but the necessary teaching experience to enable them to write with authority on the subject.

The book consists of ten chapters ranging from the simple forms of chain surveying up to the more complicated subject of setting out and measuring buildings. The various sketches are extremely clear and comprehensive, and the whole volume is easy to read and assimilate.

It is up to the standard of National Certificates and Diplomas and that required by the Institution of Structural Engineers and the Royal Institution of Chartered Surveyors. Additionally, it is a volume that can be used by any Surveyor, Architect or Engineer who desires to refresh his memory on work he has carried out in the early stages of his career.

No vast amount of mathematics is necessary to understand fully the text, and anybody with a knowledge of simple trigonometry will be able to take advantage of the information set forth.

It is a text-book in itself, neat and compact in size, and very suitable to include in any library used by students or others of the professions to which reference has been made.

FREDERICK S. SNOW, O.B.E.

Past President of the Institute of Structural Engineers,
M.I.C.E., M.I.Mech.E., M.Cons.E.,
M.I.Arb., M.Inst.W., V.P.Soc.C.E. (France).

PREFACE

THE writers' intention is to provide a text-book for those who, whilst not Land Surveyors by profession, yet require a good working knowledge of the subject. We have endeavoured to present the subject-matter simply and concisely and have avoided as far as possible the use of involved mathematics.

Although this book has a practical bias, we have kept in mind the requirements of the professional bodies concerned with the subject and of the Higher National Diplomas and Certificates in Building and Engineering and the General Certificate of Education in Surveying for the Associated Examining Board.

We wish to thank Mr. F. S. Snow for writing the foreword and reading the script; Dr. Miskin of University College, London University, for many helpful suggestions; Mr. Solomon, B.A., Lecturer in Mathematics of the L.C.C. School of Building, Brixton, for his assistance on Chapter 4, and the following Institutions for permission to reproduce their examination questions:

The Institution of Civil Engineers.

The Royal Institution of Chartered Surveyors.

The Institution of Structural Engineers.

We would emphasise that no amount of 'book knowledge' will make a surveyor—FIELDWORK IS ESSENTIAL.

W. G. CURTIN

R. F. LANE

CONTENTS

1 CHAIN SURVEYING

THIS is the simplest and oldest form of land surveying and is still extensively used. It is based upon the fact that if a triangle is set out upon the ground and the lengths of the sides measured, it may later be plotted in the drawing office to any desired scale. The only instruments required are a pair of compasses and a scale rule.

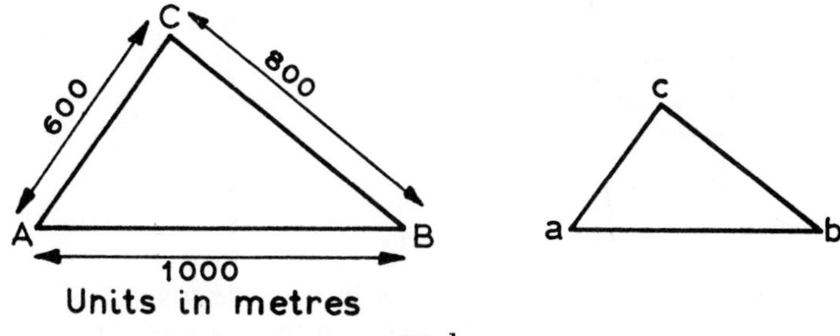

FIG. 1

Thus, if triangle ABC, fig. 1, is set out upon the ground and the lengths of the sides measured, it may later be plotted as follows:

Draw *ab* to some suitable scale, say 1 mm = 10 m, i.e. 100 mm long.

Set the compasses to the equivalent of AC, i.e. 60 mm and with centre at *a* strike an arc.

Repeat with BC with centre at *b*.

Join the intersection of the arcs *c* to *a* and *b*. The triangle thus formed will be similar to that measured on the ground, but to a scale of 1 mm = 10 m.

To survey the piece of land in fig. 2 by this method, set out a triangle in the field with its sides running as nearly parallel and as close as possible to the main boundaries of the site. Measure the sides or **'chain lines'**, and at the same time measure the **distances at right angles** from the chain line to the boundary at such places as the boundary changes direction from the line; these distances being known as **'offsets'**. From this information the shape of the field may be plotted, using the triangle as a basic framework and setting out the offsets from this.

The correct angle of short offsets is judged by eye, and they should not therefore be taken for distances over about 5 m. Where offsets of a greater length are required, other methods must be used (see page 10).

Areas other than triangular areas may be surveyed by setting up a framework of triangles and measuring offsets from them. It should be noted that triangles containing an angle of less than 30° should be

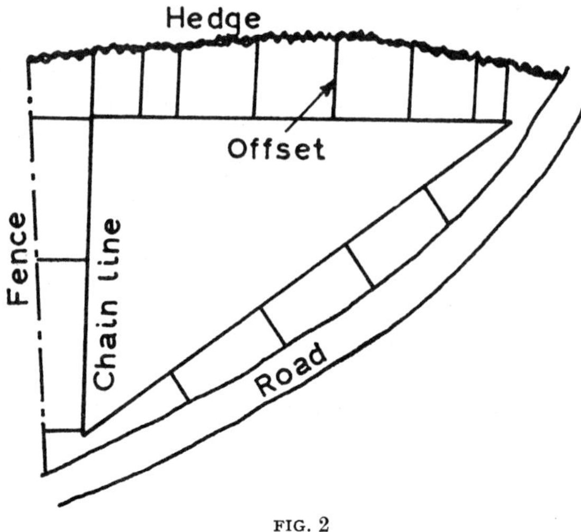

FIG. 2

avoided where possible as a small error in the measurement of one of the sides makes a considerable difference to the shape of the triangle.

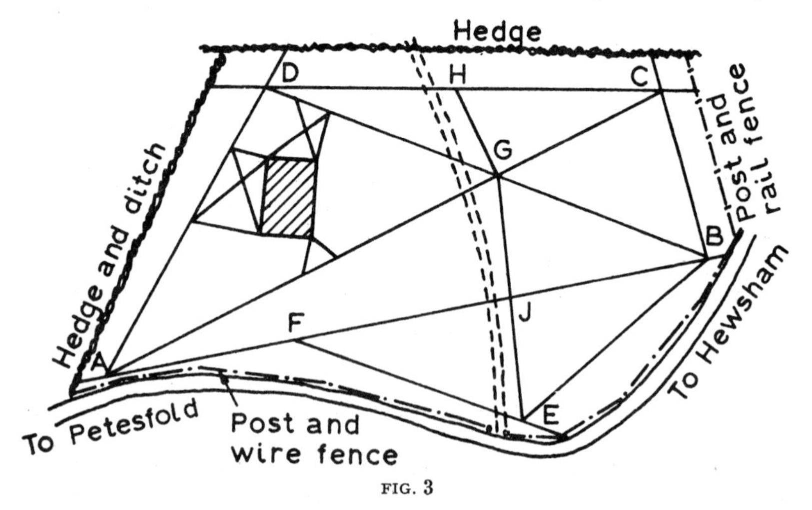

FIG. 3

Triangles of this type are known as 'ill-conditioned triangles'. The ideal triangle is one in which all the angles are of 60°, or thereabouts.

Fig. 3 shows the layout of lines to survey the area; three triangles have been used on a **base line** AC. Note that each triangle has at least one line running across it (DG, GB, JE); these are **check lines** and their purpose is to ensure that the measuring and plotting of all the lines in each particular triangle is correct, detail can also be offset from them.

When features such as buildings and roads are to be included in the survey, their positions are fixed by setting out a series of secondary lines running as closely as possible to the features from which offsets are taken. JG and GH are lines of this type.

The building shown in fig. 3 is 'tied in' to several lines by offsets which are not at right angles to the chain lines. As, however, these offsets form triangles, they can be plotted without difficulty and may also be of any desired length. The points A, B, C, D, E are known as **'stations'**, F, G, H, J as **line points.**

INSTRUMENTS

THE CHAIN—The chain is usually made of steel wire, and consists of long links joined by shorter links. It is designed for hard usage, and

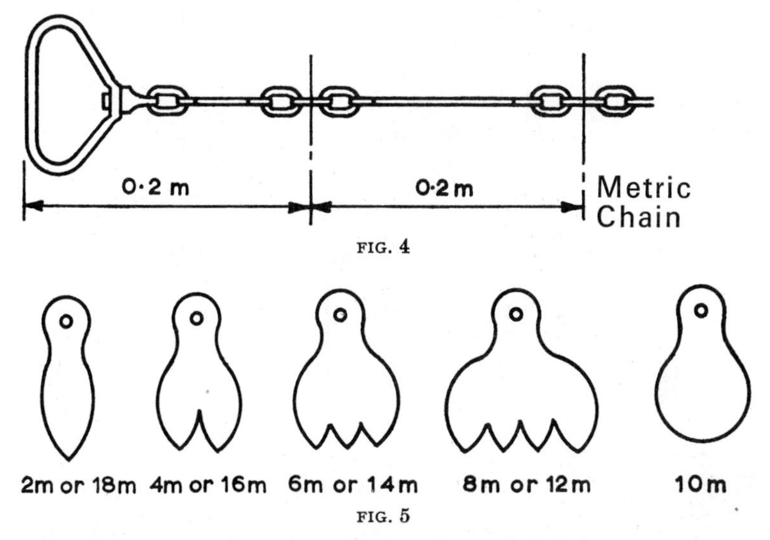

FIG. 4

FIG. 5

is sufficiently accurate for measuring the chain lines and offsets of small surveys.

The chain consists of one hundred long links, each ten links being marked by a brass tally. To avoid confusion in reading, chains are marked similarly from both ends, i.e. the tally for 2 m and 18 m is the same, so that measurements may be commenced with either end of the chain. (See figs. 4 and 5.)

Steel Band may be 30 m, 50 m or 100 m long, and 13 mm wide. It has handles similar to those on the chain and is wound on a steel cross. It is more accurate but less robust than the chain.

TAPES—Tapes are used where greater accuracy of measurement is required, such as the setting out of buildings and roads. They are marked in metres, centimetres and millimetres.

They are usually 15 or 30 m long, and are of three types:

1. Linen or linen with steel wire woven into the fabric. These tapes are liable to stretch in use and should be frequently tested for length.

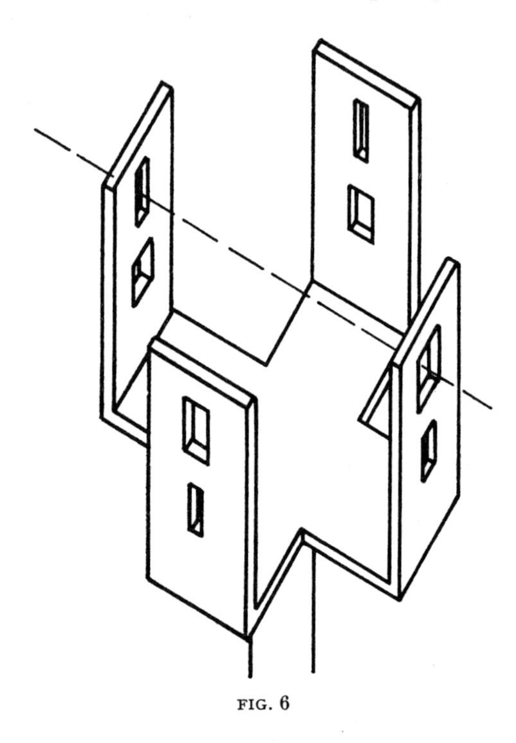

FIG. 6

They should never be used on work for which great accuracy is required.

2. Fibreglass tapes. These are much stronger than linen and will not stretch in use.

3. Steel tapes. These are much more accurate, and are usually used for setting out buildings, and always used for setting out structural steelwork.

After use in wet weather, tapes and chains should **always** be cleaned, and steel tapes dried and wiped with an oily rag.

ARROWS—Arrows consist of pieces of steel wire about $0 \cdot 5$ m long, and are used for marking temporary stations. A piece of coloured cloth is usually tied to the end of the arrow to enable it to be more easily seen in long grass.

STATION PEGS made of wood, and usually 50 mm × 50 mm and some convenient length, are used to mark the stations.

RANGING POLES are made of wood or tubular steel and are usually 2 m long. They are painted in alternate bands of red and white, the bands generally being 0·5 m deep so that the pole may be used for measuring offsets.

CROSS STAFF—This instrument is used for setting out lines at right angles to the main chain line. It consists of two pairs of vanes set at right angles to each other with a wide and a narrow slit in each vane (see fig. 6 opposite). The instrument is mounted upon a pole, so that when it is set up it is at normal eye-level.

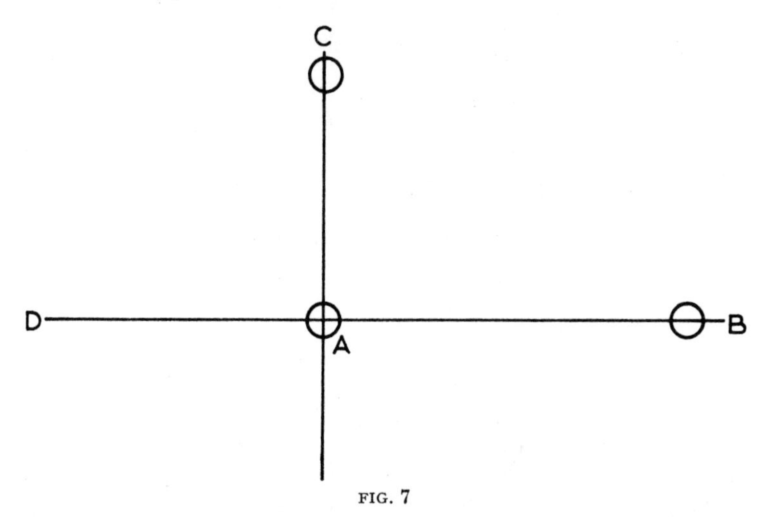

FIG. 7

Method of Use: Set up on chain line DB (see fig. 7) at point A, the point at which it is desired to set off the new line. With a ranging pole at B on line DB at some distance from A, sight the instrument upon it and then, without moving the cross staff, sight through the second pair of vanes in the direction of C and direct an assistant with a second ranging pole until this pole appears in the sights. Then line AC will be at right angles to line DB.

OPTICAL SQUARE—This instrument is used for similar purposes to the cross staff, but is easier to use and of greater accuracy. There are two forms of this instrument:

(1) the mirror method. Its construction is based on the fact that a ray of light is reflected from a mirror at the same angle as that at which it strikes the mirror.

Method of Use (see fig. 8). The instrument is held over the point C on the line AB, from which it is required to set off line CD at right angles to it. A ranging pole at B is sighted through the lower part of mirror Z, and the pole in the general direction of D is moved until its image in the upper half of mirror Z coincides with B viewed through the lower half. This line is then at right angles to line AB. Offsets up to 30 m in length may be set out in this manner.

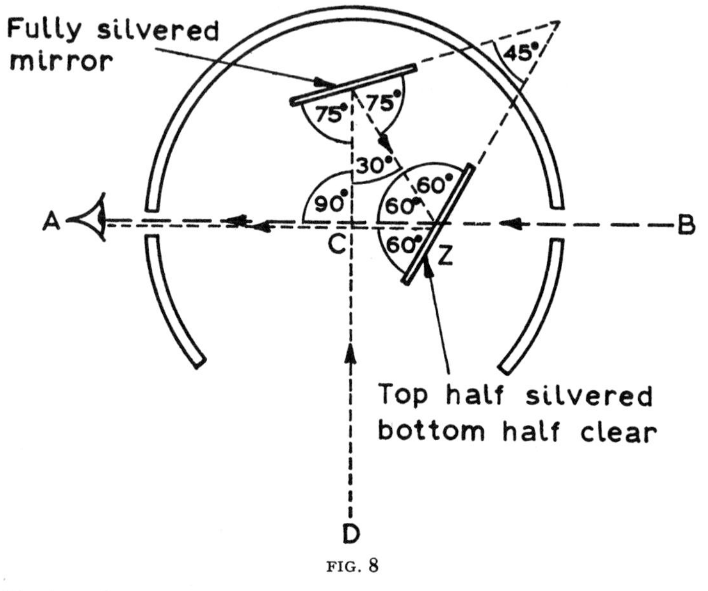

FIG. 8

(2) the prism square method. This is a simplified form of optical square consisting of a single prism (see fig. 9). The staff A is observed through a peephole above the prism, whilst staff B is reflected through the prism. No adjustment can be made to this instrument, the surveyor having to rely upon the accuracy of the prism.

WATKIN'S CLINOMETER—This instrument is used for measuring the angles of ground slopes. There are several forms, but all are basically the same. A common form is the Watkin's Clinometer, which consists of a counter-weighted scale freely suspended so that the line OX is always horizontal. The scale is divided from 0° in both elevation and depression (see fig. 10).

Method of Use: To measure the ground slope for a line AB, the surveyor stands at A whilst an assistant stands at B with a pole **clearly marked at a point at the eye-level of the surveyor**. This point is observed through the instrument at A, and if it is at a higher level than A, the instrument will be tilted upwards. As, however, the scale

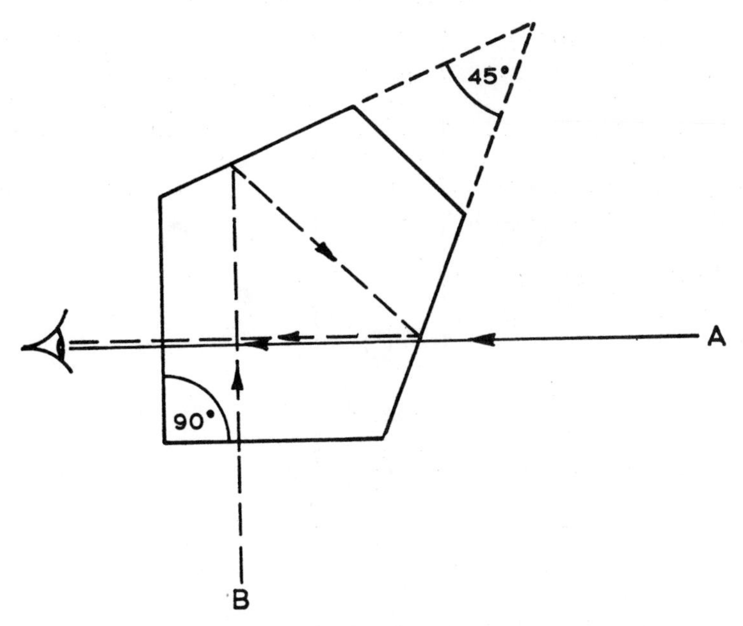
45°
90°
A
B

FIG. 9

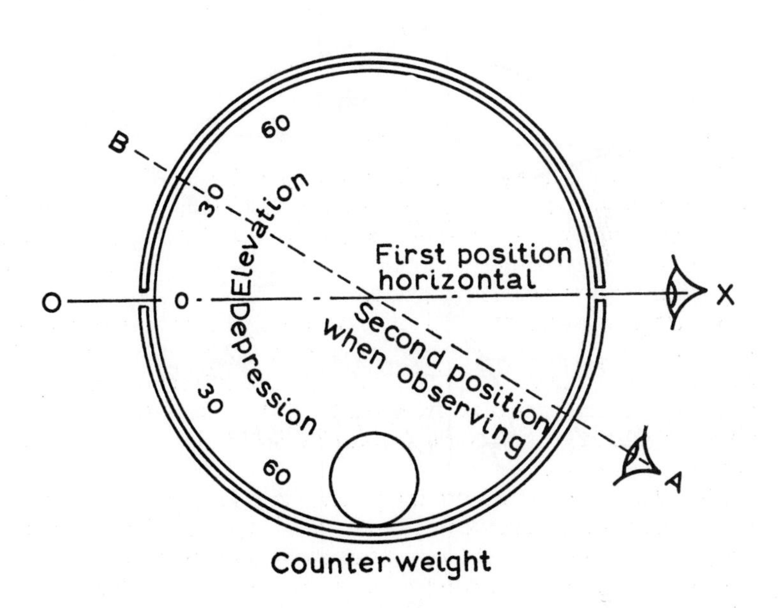
B
60
30
Elevation
First position
horizontal
O
0
X
Depression
Second position
when observing
30
60
A
Counterweight

FIG. 10

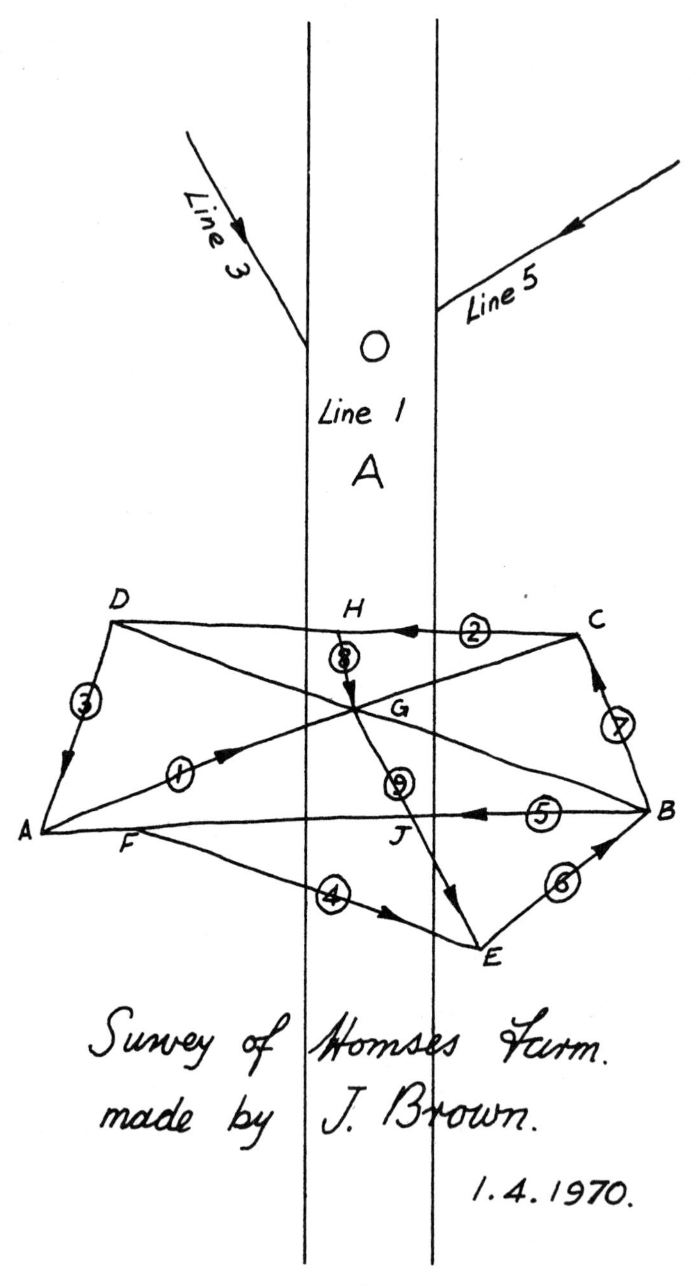
Line 3
Line 5
O
Line 1
A
D
H
C
G
A
F
J
B
E
Survey of Homses Farm.
made by J. Brown.
1.4.1970.

FIG. 11

is freely suspended, its position in relation to the horizontal will not alter and the angle of elevation may be read upon the scale from a fixed mark on the clinometer case.

ABNEY LEVEL—See page 85.

THE FIELD BOOK is about 100 mm × 200 mm and opens lengthways, the pages being ruled as in fig. 11. When using this, the entries are commenced at the **back** and continued towards the **front**. Thus the surveyor records measurements forward in the same direction as that in which he is walking. Always commence by inserting the date and place of survey, together with the name of the person making it. This should be followed by a diagrammatic sketch of the chain lines, the stations being lettered, and the number of the line and the direction in which it is to be measured shown (see fig. 11).

When entering details of the survey, the following should be noted:

1. Features such as hedges, etc., are sketched in the book.

2. No features are drawn across the pair of lines, these being used for dimensions only.

3. Where other survey lines intersect the line being measured, their direction is indicated.

4. The distance of hedges, etc., from the main chain lines are inserted beside the dimension on the line from which the offset was taken.

For illustration of above points, see back of dust cover of this book.

PROCEDURE IN MAKING A CHAIN SURVEY

1. Reconnaissance. Walk over the area to be surveyed, and note the general layout, the position of features and the shape of the area.

2. Decide upon the framework to be used and drive in the station pegs. These should, where possible, be tied in to permanent objects so that they may be easily replaced if moved (see fig. 12). At least two ties are necessary, three are preferable.

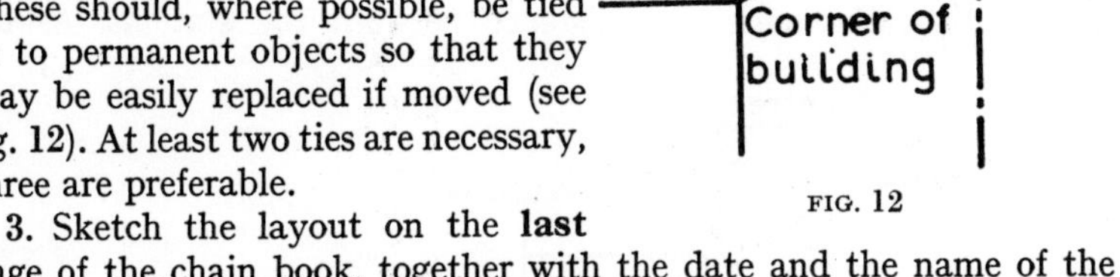

FIG. 12

3. Sketch the layout on the **last** page of the chain book, together with the date and the name of the surveyor. The longest line of the survey is usually taken as the **base line** and is measured first.

METHOD OF MEASURING A CHAIN LINE

Insert ranging poles as close as possible to the station pegs in order that the position of the peg may be located from a distance. In the case of long lines, poles are placed immediately between the stations and lined in by eye. This is done to enable a straight line to be measured. To measure line AB, the surveyor's assistant or chainman takes ten arrows and, holding one end of the chain, walks from A towards B, the surveyor remains at station A. When the chain is fully extended, the surveyor 'lines' in the chainman with station B, who then inserts an arrow at point A_1. The surveyor then walks along the chain, measuring any offsets required. Upon the completion of these measurements, the chainman moves to point A_2 dragging the chain with him, whilst the surveyor remains at A_1. When the arrow has been inserted at A_2, the surveyor removes the arrow at A_1 and proceeds to take further offset measurements. This process is repeated until the end of the line is reached or the chainman's arrows are exhausted. The collection of these arrows by the surveyor forms a check upon the number of chains measured.

METHODS OF SETTING-OUT OFFSETS TO THE CHAIN LINE

1. By eye (not to exceed 5 m in length).

2. By swinging the tape. The zero of the tape is held on the point of detail. A convenient graduation is selected on the tape, and the chainage of its intersection with the chain noted. The tape is then swung and the chainage of its second intersection noted, the required chainage being the mean of the two. The offset distance is always the shortest distance read on the tape.

3. By optical square or cross staff (see page 5) (not to exceed 30 m in length).

SLOPING SITES

In some cases the angle of slope of the ground between two stations is such that the length of the line measured on the slope is appreciably greater than the horizontal distance between the stations. As the

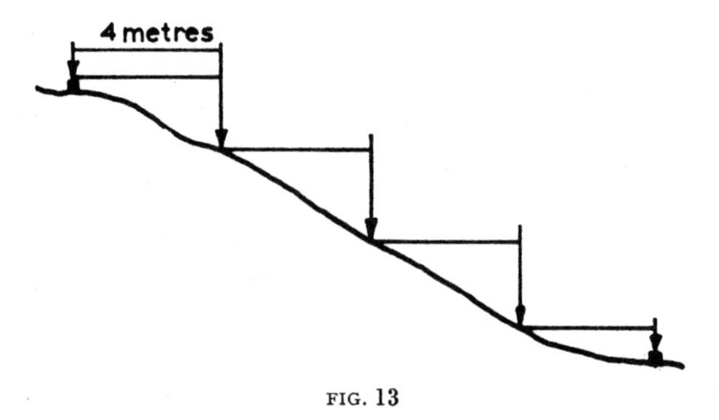

FIG. 13

horizontal distance is required for plotting the survey, any slope greater than 3° should be adjusted in one of the following ways.

(*a*) Stepping the Chain: Short lengths of chain are held horizontally and plumbed down to the ground surface, the action being repeated until the second station is reached (see fig. 13).

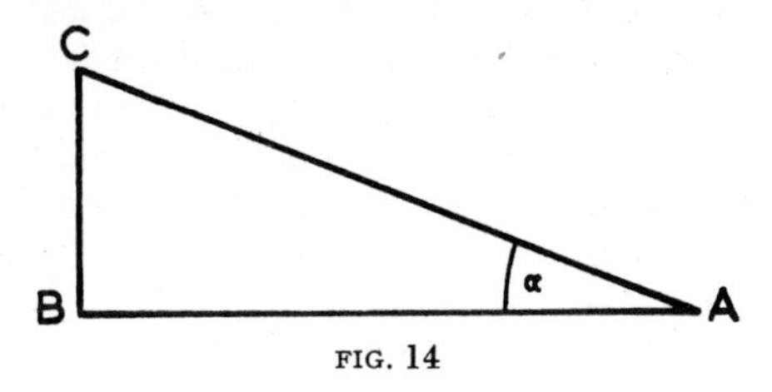

FIG. 14

(*b*) Measure the angle of slope by means of a clinometer, and calculate the horizontal length of the line by trigonometry (see fig. 14). AB = AC cos *a*.

OBSTACLES

1. TO RANGE A LINE OVER A HILL BETWEEN TWO STATIONS NOT VISIBLE FROM EACH OTHER

Set up poles at stations A and B (see fig. 15). An assistant with a ranging pole is stationed at B_1 approximately on the line AB and from where he can just see the pole at A. The surveyor with a further pole

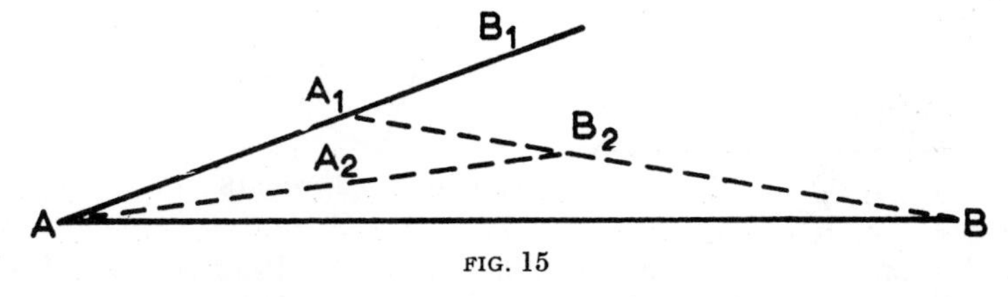

FIG. 15

walks from A towards B_1 until he can just see B. He is then lined in between A and B_1 by the assistant, and places his pole at A_1. The assistant is then lined in between A_1 and B by the surveyor, his position now being at B_2 where he can just see A. The surveyor is then lined in by his assistant to position A_2 on the line B_2A, where he can just see B. This process is continued until all poles are in a straight line.

2. SETTING OUT A RIGHT ANGLE FROM A CHAIN LINE BY THE 3, 4, 5 METHOD

This is based on the fact that a triangle having sides whose lengths are in the ratio of 3, 4, 5 is a right-angle triangle. The procedure is as

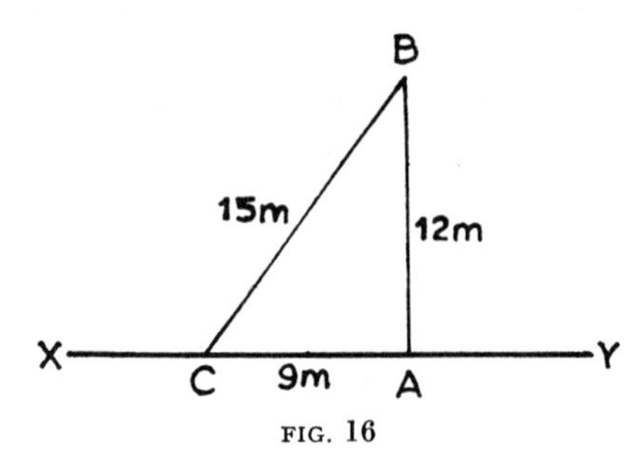

FIG. 16

follows. To set up a line AB at point A on line XY and at right angles to XY, measure back towards X a distance of 9 m to point C, then with the end of the tape held at A and the 27 m point held at C hold the tape at the 12 m point and pull taut, this will mark the point B, and AB will be at right angles to XY (see fig. 16).

3. TO MEASURE A LINE ACROSS A POND

To measure the distance BC (see fig. 17), range AB through to C. Line through B to any convenient point D, which clears the pond.

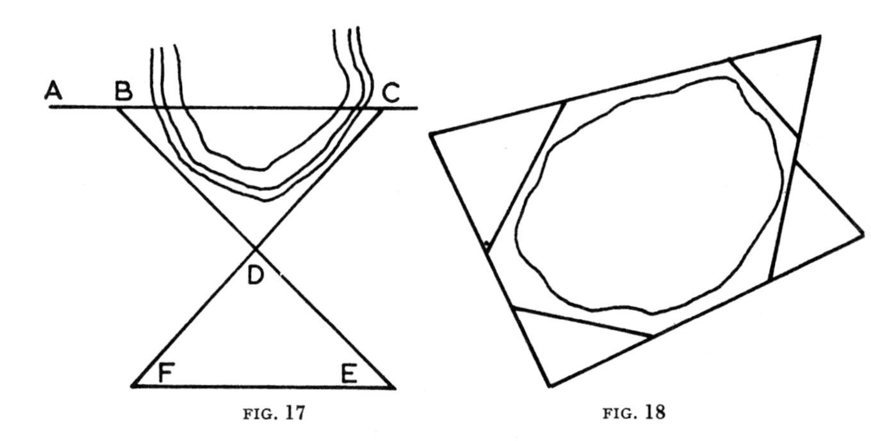

FIG. 17

FIG. 18

With a pole at D range in E making DE = BD. Produce CD to F so that DF = CD, then by identical triangles FE = BC.

4. TO SURVEY A LAKE OR THICKLY WOODED AREA

A framework of lines is set out around the area, the corners being triangulated (see fig. 18). This method is not very accurate, but often proves useful. It is known as **chain traversing.**

5. TO MEASURE THE LENGTH OF A CHAIN LINE CROSSING A RIVER

(*a*) **At right angles** (see fig. 19): Set up poles at A and B. Set out AD at right angles to the chain line. Set up a pole at C so that AC = CD. Set out DE at right angles to AD, and locate E so that

ECB is one straight line. Then ABC and CDE are identical triangles and AB = DE.

(*b*) **Obliquely** (see fig. 20): Set up poles at A and B. Locate C so that BCA is a right angle (by optical square). Continue CB to D so

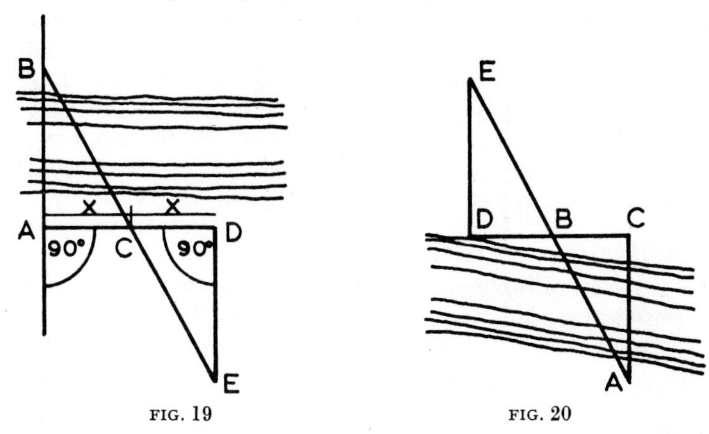

FIG. 19

FIG. 20

that DB = BC. Erect DE perpendicular to DC. Then triangle BDE = triangle BCA and AB = BE.

POSSIBLE ERRORS IN CHAINING

MISTAKES

1. Omitting to book one or more chain lengths.
2. Confusing tallies.
3. Miscounting the links.
4. Mistaking 9 m for 6 m on the tape.
5. Incorrect marking of chain end.
6. Incorrect booking.
7. Mistaking dimensions when called out.

CUMULATIVE ERRORS. ACCUMULATE IN ONE DIRECTION

Positive errors which cause the measurement to be **too large**:

1. Chain short due to: (*a*) bent links, (*b*) knots in the connecting rings.
2. Not allowing for slopes when the slope exceeds 3° per tape length.
3. Incorrect alignment.
4. Bellying out of tape due to wind.
5. Links clogged with mud.

Negative errors which cause the measurement to be **too small**:

1. Flattening of connecting rings.
2. Stretching of the tape.

The permissible error in chain surveys is 1 in 500.

COPYING, ENLARGING AND REDUCING PLANS

1. METHOD OF SQUARES

Divide the area to be plotted into a number of equal squares either by fine pencil lines drawn upon the original plan or by drawing the lines on a piece of tracing-paper and laying this over the plan. Draw a similar system of squares on the paper on which the enlarged or reduced plan is to be plotted, these squares being proportionally larger or smaller than the original squares.

Both systems of squares should be numbered, and the detail enclosed in the original square sketched in the same square in the enlarged or reduced system.

2. PROPORTIONAL COMPASSES (see fig. 21)

These consist of two legs each with a slot down the centre, connected by a pivot screw which may be loosened to allow the legs to rise within the limits of the slots. Each leg has two scales marked on it, lines

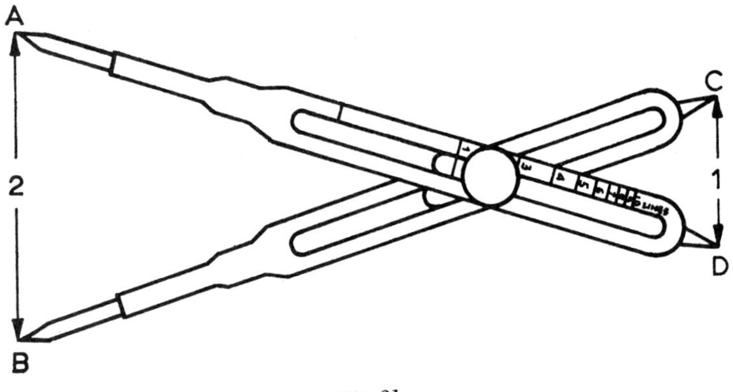

FIG. 21

and circles being on one leg and solids and plans on the other. When enlarging or reducing plans, the lines scale is used. Attached to the pivot screw are two runners marked with a line one of which slides in each slotted leg. To enlarge a plan in the proportion of, say, 1 to 2, first close the compasses and place the line on the runner against 2 on the scale, tighten the pivot screw and open the compasses. The distance AB and CD will now be in the ratio 2 to 1. An adjustable set-square is a useful aid for work containing lines at varied angles to each other.

3. TRIANGULATION METHOD

1. Draw triangle ABC on the plan, enclosing the area which is required to be enlarged or reduced.

2. Draw a similar triangle A′B′C′ to the enlarged or reduced scale required on a piece of tracing-paper.

3. Lay the tracing-paper over the plan so that A and A′ coincide and line A′B′ lies over line AB (see fig. 22A).

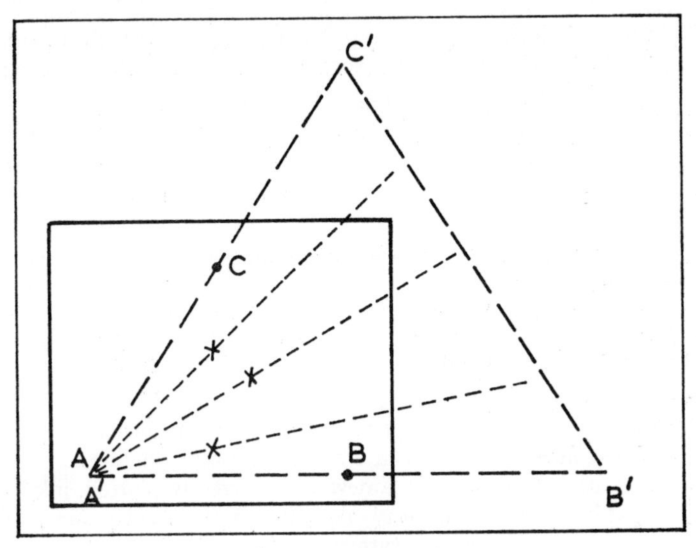

FIG. 22A

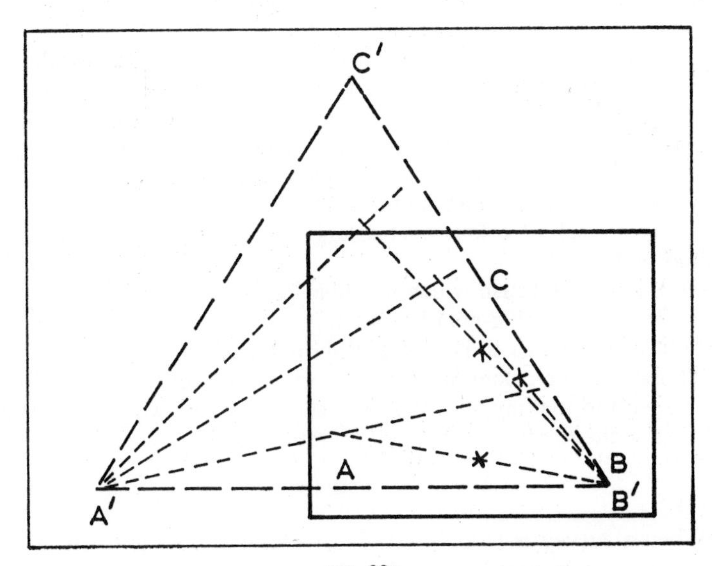

FIG. 22B

4. From A′ draw rays on the tracing-paper through the points on the plan which are required to be reproduced on the enlarged or reduced plan.

5. Move the tracing-paper so that BB′ coincide and line A′B′ lies over line AB (see fig. 22B).

6. From B′ draw rays on the tracing-paper through the same points on the plans as those rays from AA′. The intersection of the rays fixes the points on the tracing-paper to the new scale.

7. For greater accuracy, the procedure may be repeated from point C′.

4. REPLOT SURVEY

If the original survey notes are available, and time permits, it is best to replot the survey.

PLOTTING SURVEYS

1. Use good-quality paper.

2. Draw in the chain lines and tie lines in fine pencil; letter the stations, and number and indicate the direction of the lines.

3. When satisfied that the framework is correctly plotted, ink in the lines in red ink (this is not always done in practice). If the survey is to be coloured, the ink must be waterproof.

4. Plot in the detail in pencil, and when satisfied that it is correct, ink in with waterproof indian ink.

5. Draw in the north point, and magnetic north and the scale. The scale should **if possible** be as long as the longest line upon the survey.

6. Insert the title and any other lettering required. This should be in plain block capitals, elaborate forms of lettering should not be used.

7. If it is required to colour the plan, the colour washes must be very light, a common mistake being to make washes too strong.

To lay a wash of colour over a large area, the following procedure should be followed:

(*a*) Use a large brush which will hold plenty of colour.
(*b*) Tilt the drawing-board at an angle of about 30°.
(*c*) Cover the area to be coloured with a wash of clear water.
(*d*) Lay on the colour wash with horizontal strokes, working down the paper and making sure that any surplus colour which runs to the bottom of the drawing is mopped up immediately.

SCALES

When a survey is to be plotted, a suitable scale has to be decided upon. This scale may be stated as 10 mm = so many metres. This is expressed by what is known as its representative fraction, thus 10 mm = 100 m has a representative fraction of

$$\frac{10}{100 \times 1000} = \frac{1}{10\,000}$$

Common scales are, for working surveys, 1/500, and for preliminary surveys 1/1250.

DIAGONAL SCALE

This type of scale enables very small distances to be measured. It is constructed as follows:

To set up a scale of 1 mm = 1 m to measure to 0·05 m:

1. Set out a line 100 mm long and divide into 10 equal parts.
2. Set out a line of a convenient length at right angles and at one end of the last line, and divide into 20 equal parts.
3. Complete the scale as shown in fig. 23.

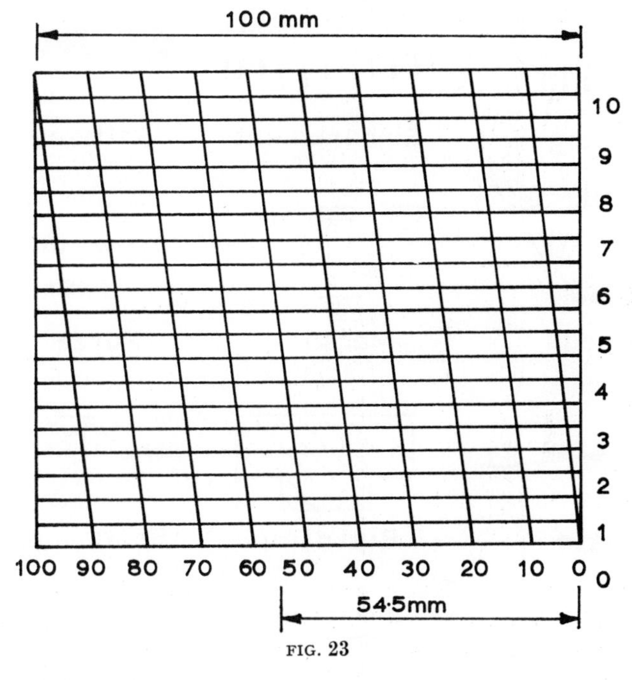

FIG. 23

CONVENTIONAL SIGNS

There are a considerable number of these, most of which can be seen on Ordnance maps. The signs used for certain features on large-scale maps sometimes differ from those used on the smaller scales. Those shown in fig. 24 are the more common ones used for large-scale surveys.

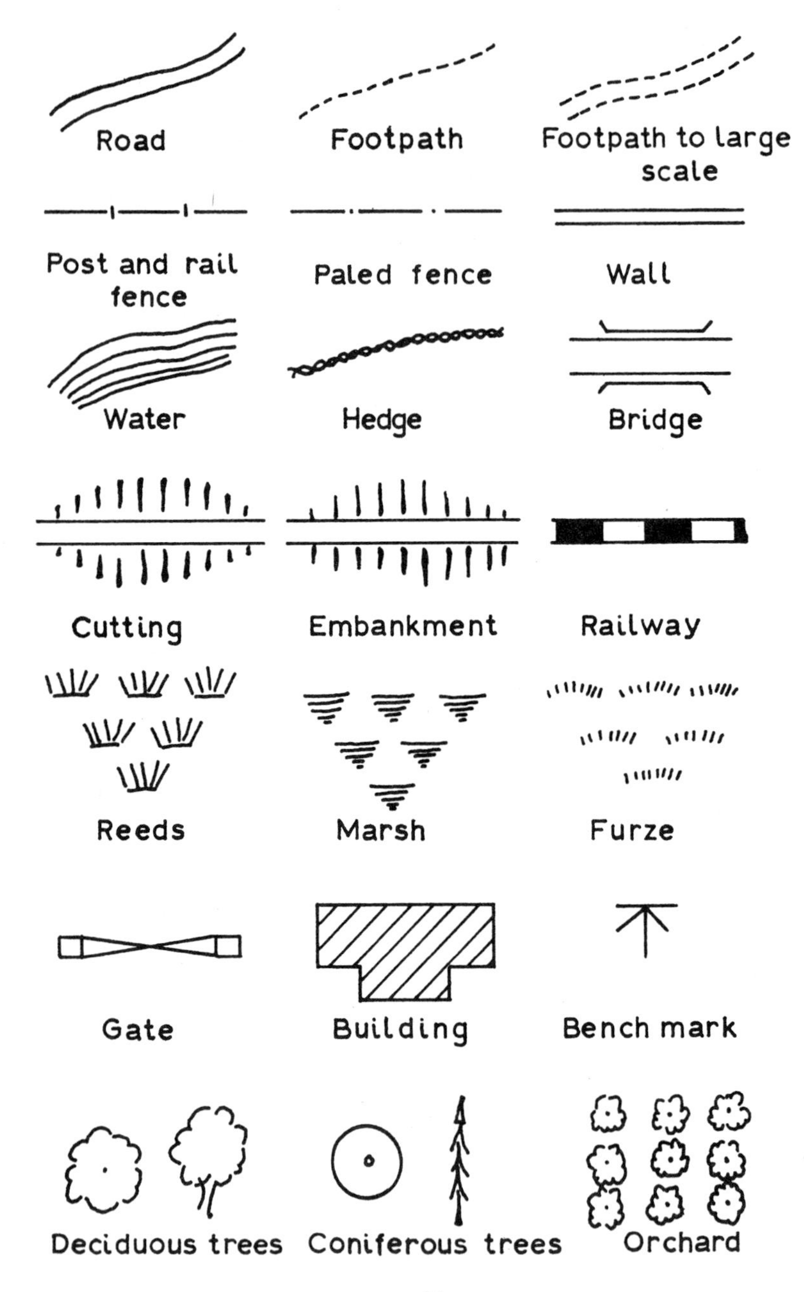
Road
Footpath
Footpath to large scale
Post and rail fence
Paled fence
Wall
Water
Hedge
Bridge
Cutting
Embankment
Railway
Reeds
Marsh
Furze
Gate
Building
Bench mark
Deciduous trees
Coniferous trees
Orchard

FIG. 24

Examples

1. Construct a scale of 1 in 500 to read in metres to 1 m and long enough to show 100 m.

Solution

$$\text{Length of scale} = \frac{\text{Total length to be measured}}{\text{Rate per metre}} = \frac{100\ \text{m}}{500} = 0{\cdot}2\ \text{m long.}$$

Mark off a line 0·2 m long and divide into 10 equal parts. Subdivide the first part into 10, then each part will represent 1 m.

2. The plan of a field of an area of 17·436 hectares covers 27 900 mm² of paper. What is the scale?

Solution

$$\begin{aligned} 1\ \text{hectare} &= 100\ \text{m} \times 100\ \text{m} \\ &= 10\ 000\ \text{m}^2 \\ 17{\cdot}436\ \text{hectares} &= 174\ 360\ \text{m}^2 \\ \therefore\quad 1\ \text{m on paper} &= \sqrt{\frac{174\ 360}{0{\cdot}0279}}\ \text{metres in the field} \\ &= \sqrt{6\ 250\ 000}\ \text{metres} \\ &= 2500\ \text{metres} \end{aligned}$$

Therefore the scale is 1 in 2500.

3. On a plan of scale 1 in 600, the distance between two points was measured and found to be 428 m. It was afterwards found that the scale used was one of 1 in 500. What was the true length?

Solution

For every metre measured 500 m only were credited, whereas 600 m was the actual distance covered, i.e. 1·2 times as much. Actual length of line = 1·2 × 428 = 513·6 m.

4. A field measured with a 20 metre chain was found to be 10 hectares. The chain was later found to be 19·7 m long. What is the true area of the field?

Solution

$$\frac{\text{True area}}{\text{Calculated area}} = \frac{(\text{Actual length of chain})^2}{(\text{Nominal length of chain})^2}\,.$$

$$\begin{aligned} \text{True area} &= \frac{19{\cdot}7^2 \times 10}{20^2} \\ &= \mathbf{9{\cdot}702\ hectares.} \end{aligned}$$

2 LEVELLING

WATER-LEVEL—An early form of levelling instrument, consisting of a glass U-tube filled with water, the tube being rigidly fixed to a stand (see fig. 25).

Line of Collimation

An imaginary line drawn between the two columns of water is horizontal, and may be visually projected beyond the instrument. This line is called the Line of Collimation.

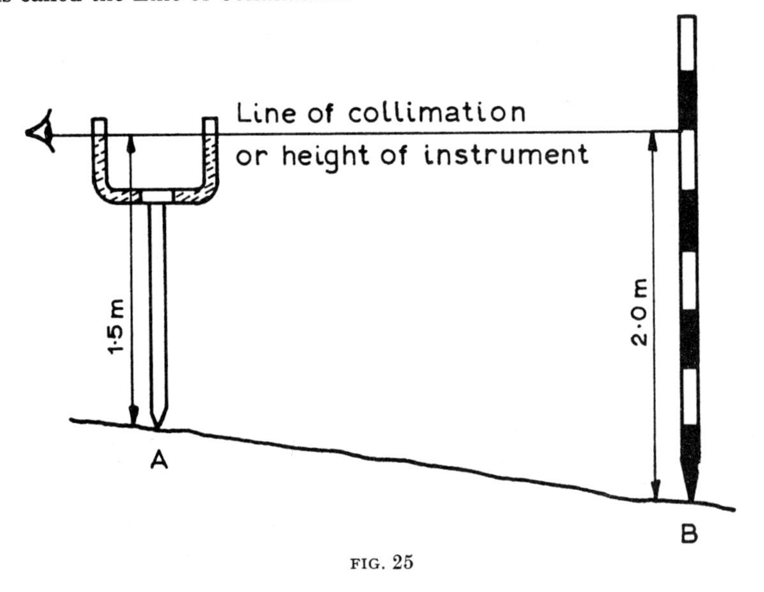

FIG. 25

APPLICATION—The height from the ground at the foot of the staff at point A to the surface of the water may be measured. In this case it is 1·5 m.

A graduated rod is held at point B, and upon looking along the line of collimation it is seen that the line intersects the staff at the 2·0 m mark.

The difference of level at the two points A and B is:

Height recorded on graduated staff — Height of instrument,
i.e. 2·0 — 1·5 = 0·5 m.

BRICKLAYER'S LEVEL—The most simple form of level is that used by bricklayers. It consists of a glass tube filled with liquid which contains an air bubble (see fig. 26).

This tube is set in a wooden block in such a way that when the instrument is placed upon a horizontal surface, the bubble floats centrally in the tube.

This instrument, in conjunction with a long, straight wooden lath, may be used to set a series of pegs to that their tops are all at the same level, i.e. in a horizontal line. This is carried out as follows:

Peg A is driven so that its top is at the required level. Peg B is then driven and the lath is rested upon pegs A and B. The level is placed upon the lath, and peg B is gradually driven down until the bubble floats centrally in the glass tube of the level. The tops of pegs A and B are now in the same horizontal line. The method is then repeated with pegs C and D.

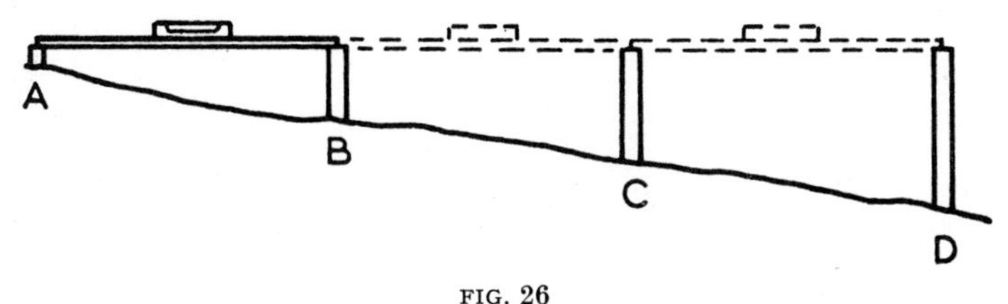

FIG. 26

THE BUBBLE TUBE—This is constructed of glass and is circular in cross section, the longitudinal section being a circular arc. The tube is filled with a liquid of low viscosity, i.e. alcohol, sulphuric acid, etc., with a small air bubble formed in it. To enable the bubble to be centred, a number of gradations are etched upon the top surface of the tube, usually 2 mm apart. The tube is mounted in a metal casing with capstan-headed screws at one or both ends. It can be adjusted by means of these, so that when the base of the mounting is horizontal, the bubble floats centrally (see fig. 27).

Thus, when the bubble floats centrally, the tangent of the tube is parallel to the base plate. The tangent of the tube is also called the **axis of the bubble**.

SENSITIVITY OF THE BUBBLE

This depends upon the radius of the curvature of the tube. The greater the radius the more sensitive is the bubble. The radius is generally between 20 m and 30 m.

The use of the water-level and the bricklayer's level is limited by the distance that can be observed with the naked eye. It was found necessary, therefore, to attach a telescope to the latter instrument in order to increase the distance of the sight.

TELESCOPE LENSES

Types of Lenses (see fig. 28): The surfaces of all lenses are of spherical form.

Rays *xx* passing through the principal axis, or optical centre, are not deflected. All other rays are refracted. D is the distance from the lens centre to principal focus F and is known as the focal length.

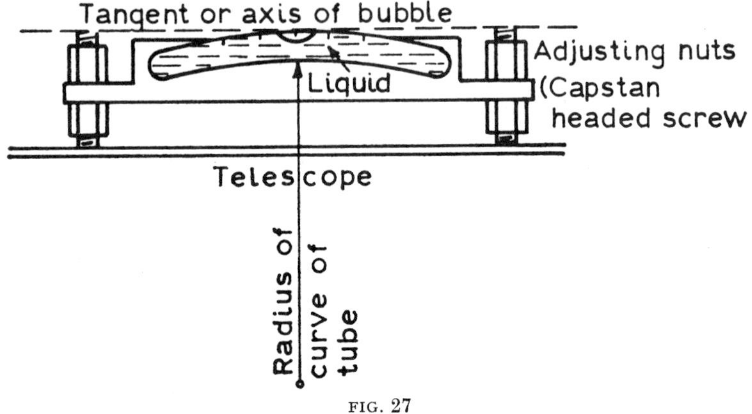

FIG. 27

MAGNIFICATION

The double-convex lens, the type used as a 'Reading Glass', functions as follows:

The object AB, which is actually a number of points radiating cones of light rays, is held close to the lens; one ray from B parallel to the

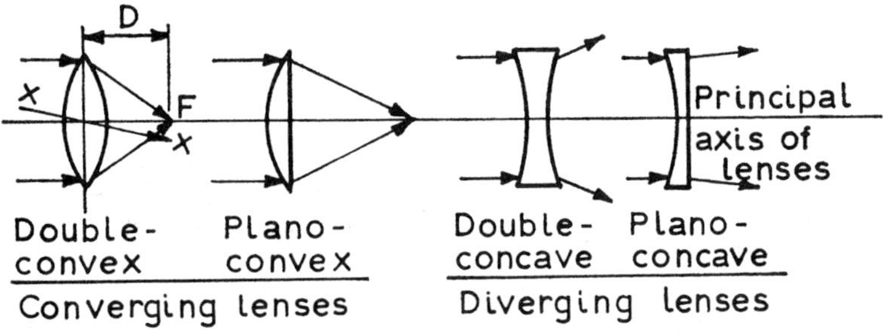

FIG. 28

principal axis of the lens strikes the lens at C. It is refracted through the focal point at D to the eye. The ray BO passing through the optical centre is not refracted. Both rays appear to come from B′, and thus B′A′ appears larger than BA (see fig. 29).

THE TELESCOPE is a magnifying instrument working upon the same principle as the reading-glass, but modified to view distant objects.

The modification consists of an object lens which collects light rays from the object and forms an inverted image in front of the magnifying lens (see fig. 30). A double-convex lens is used as the object glass, and has the effect of reversing the image, which appears to the observer to be inverted.

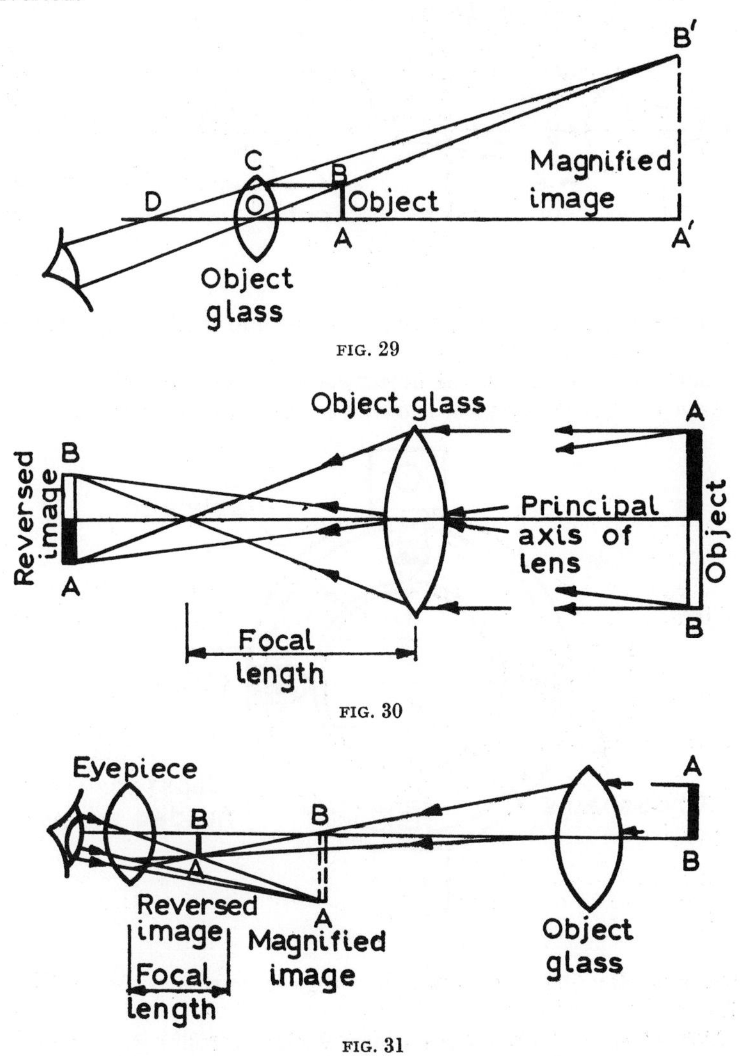

FIG. 29

FIG. 30

FIG. 31

THE EYEPIECE—When a double-convex lens is used as the eyepiece, the magnification is developed as shown in fig. 31.

KEPLER'S TELESCOPE is the form most used in surveying instruments. It comprises two convex lenses on the same principal axis, that nearest the object being known as the object glass whose function it is to collect

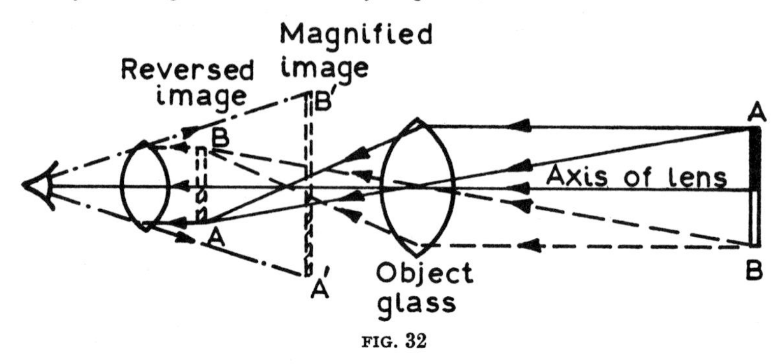

FIG. 32

the light rays from the object and focus them in front of the other lens or eyepiece (see fig. 32).

Light rays impinge on the object glass and focus before arriving at the eyepiece. If the eyepiece is placed so that AB is in the focal length,

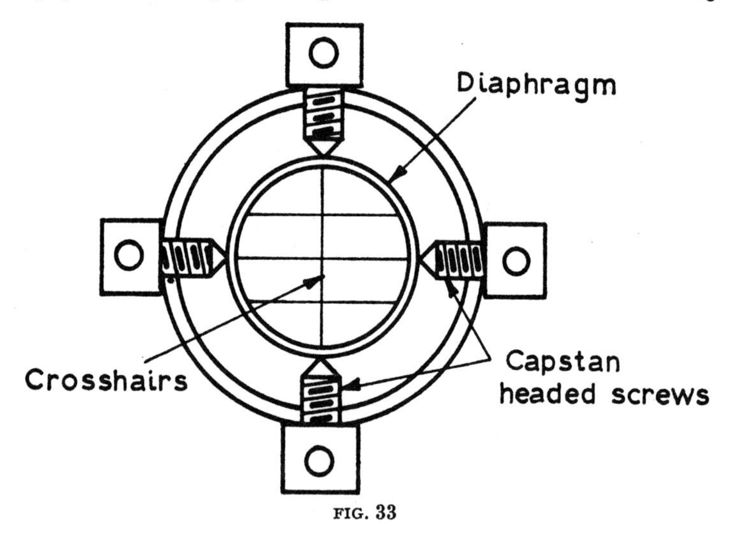

FIG. 33

the rays, after refraction from the eyepiece, appear to proceed from B′A′.

Note—The axis of the lenses is also the line of collimation.

DIAPHRAGM—This consists of a metal ring with fine cross hairs mounted on it. These hairs are etched upon a piece of glass (see fig. 33). The diaphragm is placed in the plane of the reversed image AB (see fig. 32), and is magnified with it. The cross hairs are so placed as to intersect upon the line of collimation, and thus enable the exact point at which this line cuts the graduated staff to be read with ease. In order that the diaphragm may be adjusted, it is held by capstan-headed screws.

EYEPIECE—The single lens previously mentioned is usually replaced by an eyepiece composed of two plano-convex lenses (see fig. 34). This

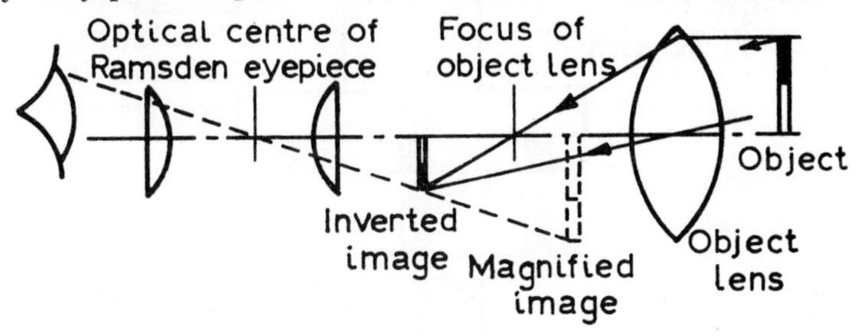

FIG. 34

is known as the Ramsden Eyepiece. It magnifies the image but does not alter its inverted position.

INTERNAL FOCUSING

This is obtained by inserting a lens between the object glass and the eyepiece. This lens is moved to and fro along the tube by a ratchet, and thus alters the direction of the light rays (see fig. 35).

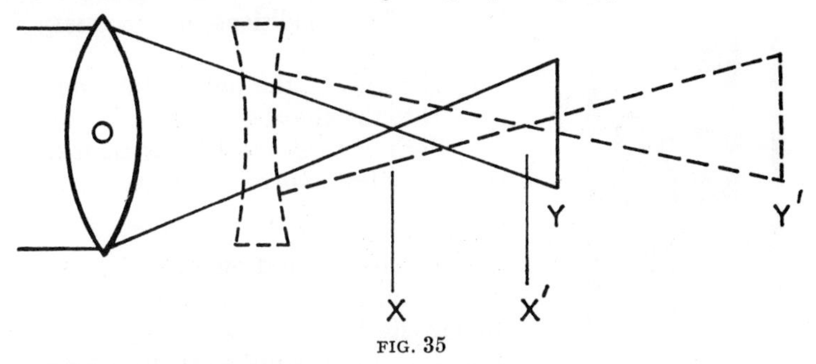

FIG. 35

Rays are refracted through O to X, but when a double-concave lens is inserted it alters the path of the rays which focus at X', and thus move the image from Y to Y'.

DUMPY LEVEL

This is the most common form of the surveyors' level and consists of a bubble tube attached to a telescope. The telescope is rigidly fixed to a stage which has three or four foot-screws. These foot-screws are

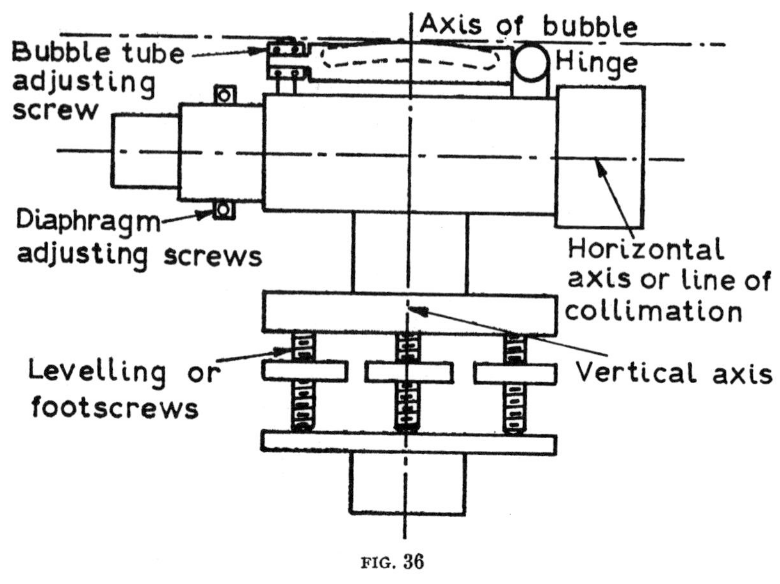

FIG. 36

FIG. 37

used to bring the bubble to the centre of its run, and thus make the axis of the bubble horizontal. Then, if the line of collimation is parallel to the bubble axis, the line of collimation is also horizontal (see fig. 36).

Note—Two conditions are necessary for accurate work.

1. The axis of the bubble tube must be parallel to the line of collimation.
2. Both of these must be at right angles to the vertical axis of the instrument.

Levelling staff

The graduated rod mentioned previously usually takes the form of a telescopic or folding staff 4·25 or 5 m long. The most common form is rectangular in section and in three lengths The face of the staff is graduated in metres, decimetres and centimetres as shown in fig. 37. It is read to millimetres by estimation

SETTING-UP THE LEVEL

This is sometimes referred to as 'temporary adjustment'.

1. Open tripod legs to about 60° and press firmly into the ground.

2. Open box and note carefully how the instrument is packed—neglecting to do this may cause considerable difficulty when replacing the instrument, which if done incorrectly is a source of much of the damage caused to surveying instruments.

3. Lift instrument from box, **but never by the telescope tube.**

4. Screw firmly on to the tripod, taking care not to cross the threads.

5. Roughly level the instrument by adjusting the tripod and complete by means of the foot-screws, i.e. make the bubble axis at right angles to the vertical axis of the instrument, as follows:

3-SCREW LEVEL

Position 1, telescope parallel to screws A and B: Turn screws A and B in opposite directions until the bubble is in the centre of the run (the bubble will follow the movement of the left thumb).

Position 2, at right angles to position 1: Turn screw C until bubble is in the centre of the run.

Position 3: Turn back to position 1 and repeat until bubble remains central when telescope is turned in any direction (see fig. 38).

4-SCREW LEVEL

Position 1: Turn screws A and D.

Position 2: Turn screws B and C.

Position 3: Turn back to position 1 and repeat until bubble

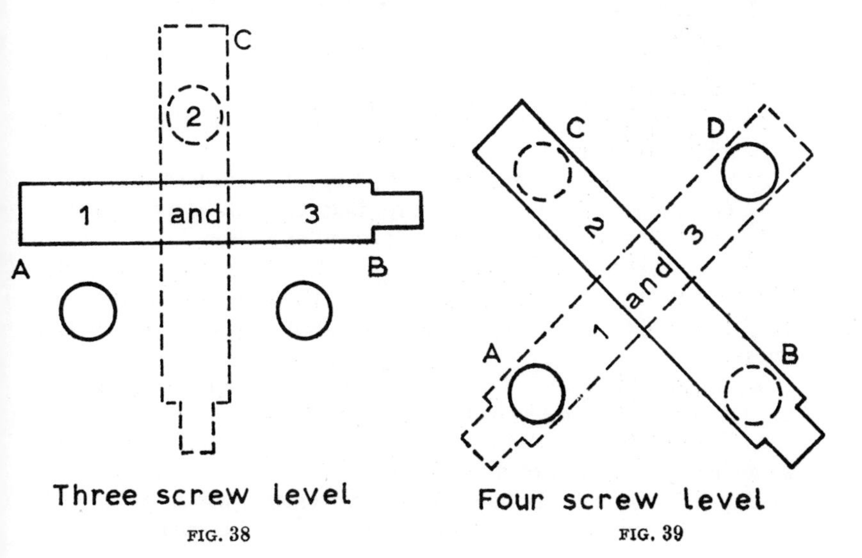

FIG. 38

FIG. 39

remains central when telescope is turned in any direction (see fig. 39).

6. Focus cross hairs. This is best done by putting the telescope **out** of focus and focusing the cross hairs by moving the eyepiece.

7. Sight on the staff and focus the telescope.

If steps 6 and 7 are not carried out in the above order, the focus of the staff may not be exactly in coincidence with the cross hairs. This is known as **parallax**, and the surveyor, by moving his eye up and down, would see the cross hairs moving up and down, and thus obtain different and inaccurate readings on the staff.

TERMS USED IN LEVELLING

Bench Mark (B.M.): A fixed point on the earth's surface whose level above Ordnance datum is known.

Ordnance Datum (O.D.): Mean sea-level, to which all other levels are related. Laid down by the Ordnance Survey.

Backsight: The first sight taken after the level has been set up. A sight taken to a point whose height is known or can be calculated.

Foresight: The last sight taken. A sight taken to a point whose height is required to carry on the line of levels.

Intermediate Sight: Any other sight taken.

Reduced Level (R.L.): Calculated level of a point above or below the datum.

Height of Instrument (H. of I.): The height of the line of collimation above the datum.

Change Point: The point at which both a foresight and then a backsight are taken.

METHOD OF USE

To find the difference in level between points A and F (see fig. 40):

1. Set up the instrument at station 1.
2. Hold the staff at A and take a reading; this will be a backsight.
3. Transfer staff to B and take a reading; this will be a foresight.
4. Leave staff at B and set up instrument at station 2.
5. Take a second reading on the staff at B; this will be a backsight.

The lengths of the backsight and foresight at any level station should be approximately equal to minimise the effect of possible instrument errors.

6. Transfer the staff to C and take a reading: The reading taken on to D is an intermediate sight from level position 3.

7. The above process is repeated until a reading at F is obtained. As a check on the staff reading, the first and last points must be bench marks. The check is obtained by the agreement, within permissible error, of the reduced level of the last point with its known level.

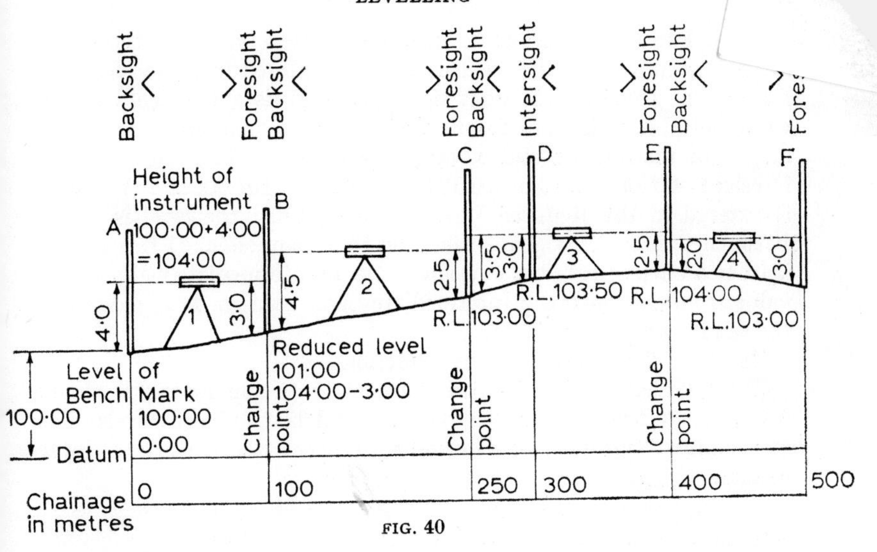

FIG. 40

LEVEL OR FIELD BOOK

There are two methods of reducing level readings, both being booked in the same way: In columns for backsight, foresight or intermediate sight, on a different line for each staff position.

1. COLLIMATION METHOD OR HEIGHT OF INSTRUMENT

Booking and reducing the levels shown in fig. 40 is carried out thus:

The first reading, 4·00, is a backsight, and is entered in the Backsight column. The sight is taken on to A, which is a Bench Mark of a reduced level of 100·00; this reduced level is entered in the Reduced Level

Back-sight	*Inter. Sight*	*Fore-sight*	*Height of Inst.*	*Reduced Level*	*Distance*	*Remarks*
4·000			104·000	100·000	0·00	A. Bench mark
		3·000		101·000	100·00	B.
4·500			105·500			
		2·500		103·000	250·00	C.
3·500			106·500			
	3·000			103·500	300·00	D. Manhole
		2·500		104·000	400·00	E.
2·000			106·000			
		3·000		103·000	500·00	F. Bench mark
14·000		11·000		100·000		
11·000				3·000		
3·000						

column. The sum of the backsight and reduced level give the height of the instrument as 104·00; this figure is entered into the Height of Instrument column. The next sight, on to B, is a foresight, and its value, 3·00, is entered in the Foresight column. The difference of the height of instrument and foresight gives the reduced level of the foresight station. This will be 104·00 − 3·00 = 101·00, and this figure is entered in the Reduced Level column. When the level is set up between C and E, a sight is taken on to a manhole at D this sight is neither a backsight nor a foresight, it is an intermediate sight and its value is entered into the Intermediate Sight column. It is reduced in the same way as a foresight.

To check the accuracy of the reduction:

The difference of the sum of the backsights and foresights should equal the difference between the first and last reduced levels. This method is quicker to reduce but does not check the reduction of intermediate sights.

2. RISE AND FALL METHOD

The terms 'rise' and 'fall' mean the rise or fall of a station relative to the preceding station. A rise will occur when the staff reading is less than the reading on the preceding station. The reading on B is 1·00 less than that on A, therefore B has a rise, relative to A, of 1·00. A fall will occur when the staff reading is more than the reading on the preceding station. The reading on F is 1·00 more than that on the preceding station, E, therefore F has a fall, relative to E, of 1·00. The rise or fall of a station is added to the reduced level of the preceding station to obtain the reduced level of the station, a fall is subtracted.

Back-sight	*Inter. Sight*	*Fore-sight*	*Rise*	*Fall*	*Reduced Level*	*Dis-tance*	*Remarks*
4·000					100·000	0·00	A. Bench mark
		3·000	1·000		101·000	100·00	B.
4·500							
		2·500	2·000		103·000	250·00	C.
3·500							
	3·000		0·500		103·500	300·00	D. Manhole
		2·500	0·500		104·000	400·00	E.
2·000							
		3·000		1·000	103·000	500·00	F. Bench mark
14·000		11·000	4·000	2·000	110·000		
11·000			1·000				
3·000			3·000		3·000		

To Check the Reduction: The difference of the sums of the backsights and foresights should equal the difference of the sums of the rises and falls and the difference between the first and last reduced level. It will be noted that in this method the intermediate sights are checked, since they affect the Rise and Fall columns, whereas in the collimation method only the backsights and foresights are checked. If it is not required to relate a survey to the Ordnance Survey, an arbitrary datum may be used, i.e. the edge of the curb, the top of a step and any convenient level, for instance 10·00 m or 100·00 m may be given to it.

When a considerable number of readings are taken, several pages of the level book will be used. To avoid laborious casting, each page may

Back-sight	*Inter. Sight*	*Fore-sight*	*Height of Inst.*	*Reduced Level*	*Distance*	*Remarks*
4·000			104·000	100·000	0·00	A. Bench mark
		3·000		101·000	100·00	B.
4·500			105·000			
		2·500		103·000	250·00	C.
3·500			106·500			
		3·000		103·500	300·00	D. Manhole
12·000		8·500		100·000		
8·500				3·500		
3·500						

END OF PAGE

BEGINNING OF NEW PAGE

Back-sight	*Inter. sight*	*Fore-sight*	*Height of Inst.*	*Reduced Level*	*Distance*	*Remarks*
3·000			106·500	103·500	300·00	D. Manhole
		2·500		104·000	400·00	E.
2·000			106·000			
		3·000		103·000	500·00	F. Bench mark
5·000		5·500		103·500		
		5·000		0·500		
		0·500				

be checked separately if it begins with a backsight and ends with a foresight. If an intermediate sight comes at the end of a page, enter as a foresight on that page and as backsight on the next page.

Note—Some surveyors prefer to put the backsight and foresight readings on a staff at the same station on the same line of the level book. This method is shown in the examples on pp. 40 and 43.

PERMANENT ADJUSTMENT OF THE DUMPY LEVEL

Fig. 41 shows a level out of adjustment. To bring the instrument into adjustment, it is first necessary to make the bubble axis at right angles to the vertical axis, and then to set the line of collimation parallel to the bubble axis.

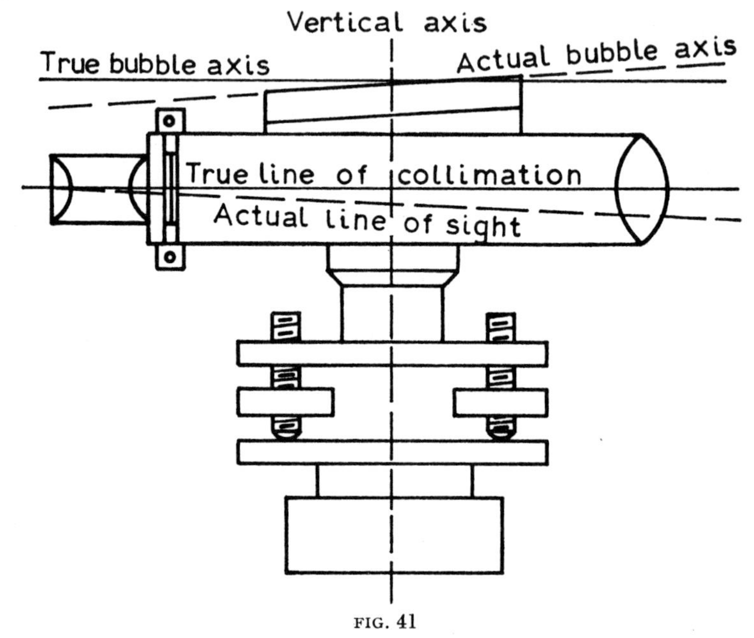

FIG. 41

ADJUSTMENT NO. 1

Bubble axis at right angles to the vertical axis

Set up and level the instrument. Turn telescope over pair of foot-screws and centre the bubble exactly. Then reverse tube, i.e. turn

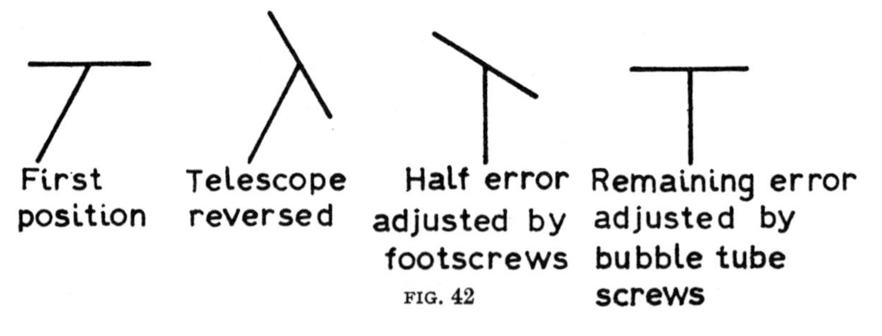

FIG. 42

through 180°. If the bubble runs off centre, bring it half-way back with the foot-screws, and the remainder of the way with the capstan-headed adjusting screw on the bubble tube (see fig. 42).

ADJUSTMENT NO. 2

Line of collimation parallel to the bubble axis

1. Set up midway between two pegs 200·000 apart. Hold the staff on one peg and note reading—say it is 5·600.

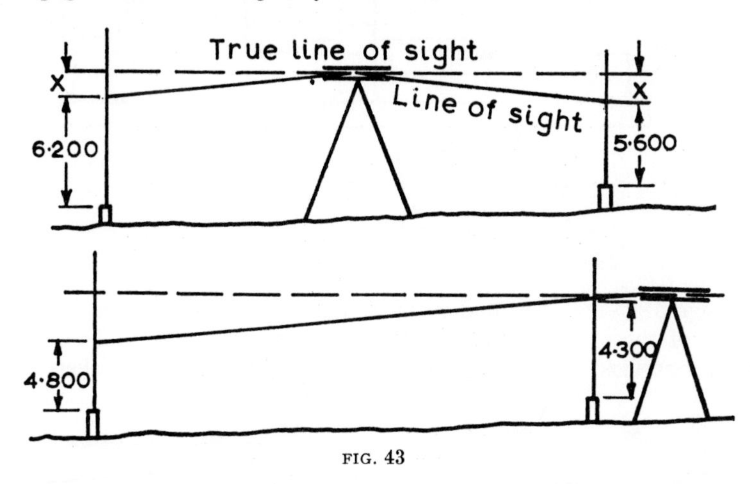

FIG. 43

2. Place the staff on the second peg and note the reading—say it is 6·200. Despite the fact that the instrument is out of adjustment, the difference in height between the pegs is 6·200 — 5·600 = 0·600, since the error in staff reading X is the same in each case.

3. Move the instrument close to one peg and again read the two staves, and obtain the difference in height 4·800 — 4·300 = 0·500. This is not the same as the true difference in height and the instrument is in error by 0·100 in the distance.

4. To adjust the error, the correct reading on the far staff must be calculated. Since the error on the near staff is very small, this reading (4·300) is accepted and the true difference in height, 0·600, is applied to it. Correct reading on far staff 4·300 + 0·600 = 4·900. The diaphragm is moved using the capstan headed screws until the reading is noted on the far staff, taking care that the bubble is central at the same time.

THE WYE LEVEL

This instrument has the advantage of being more accurate, in principle, than the Dumpy Level. It differs from the Dumpy in that the telescope tube is attached to the instrument by two Y-shaped supports and may be removed at will. It may also be revolved in the Y brackets through 360° (see fig. 44). The advantage of this is that should the line of collimation be out of adjustment, the instrument may be revolved

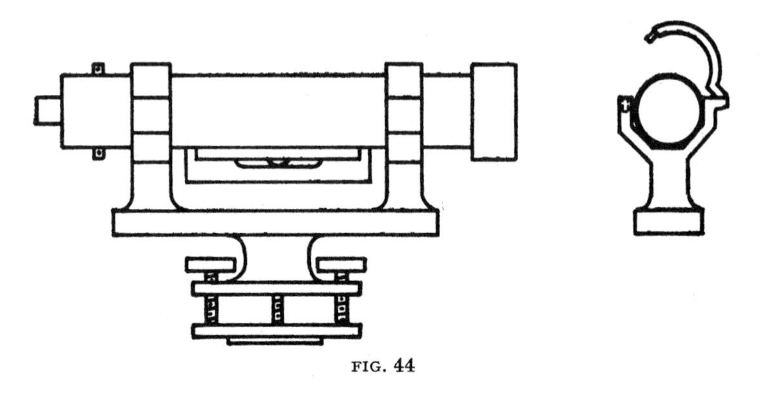

FIG. 44

through 180° and a second reading taken (see fig. 45, 1 and 2). The mean of these two readings is the true reading required. To obtain greater accuracy, the telescope may be reversed in the Y brackets (see

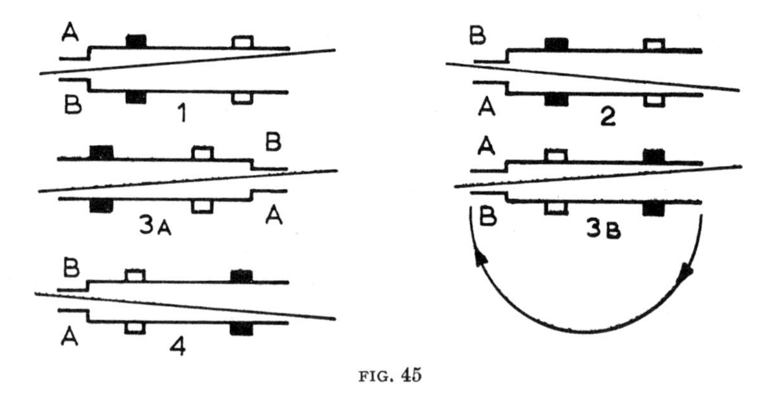

FIG. 45

fig. 45, 3A) swung through 180° horizontal (see fig. 45, 3B), and the two further readings taken (see fig. 45, 3B and 4).

The instrument is now almost obsolete.

TILTING LEVEL

An improvement on the Dumpy Level. Instead of being fixed rigidly to the stage, it is hinged so that the telescope and bubble tube may be tilted. The tilting movement is obtained by a finely threaded screw (see fig. 46), with a coil spring opposing it to ensure that the telescope is held firmly.

When using this instrument, it is only necessary to set it roughly level using a circular bubble, as the bubble is finally levelled for each reading by means of the adjusting screw, and thus greater accuracy is obtained, since the bubble axis is always made to be at right angles with the vertical axis.

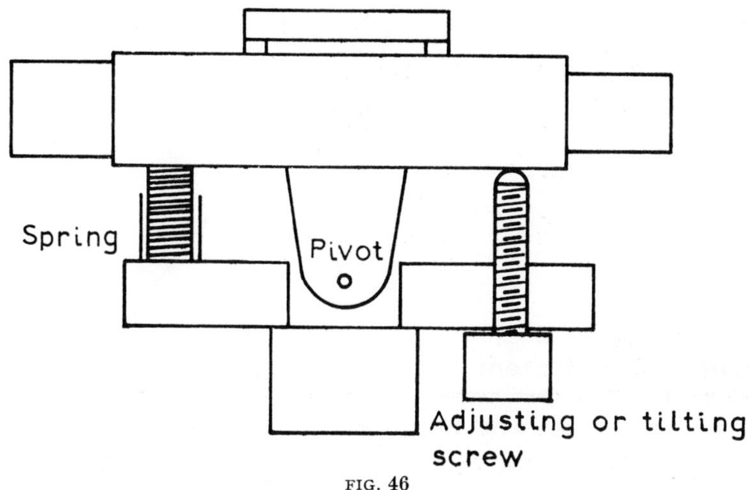

FIG. 46

The QUICKSET LEVEL is a particular type of tilting level where the three foot-screws are replaced by a ball and socket joint for quick levelling up using the circular bubble.

The tilting level has only one permanent adjustment—to set the line of collimation horizontal when the bubble is central. This is carried out in the same way as the dumpy level adjustment, using two pegs until the correct reading on the far staff is obtained. Then the horizontal hair is set on this reading using the tilting screw, and the bubble, which will then be off centre, is exactly centred using the bubble adjusting screw.

COWLEY LEVEL

This instrument, produced by Messrs Hilger & Watts, differs from other forms of level in that neither lenses nor a bubble tube are used in its construction. Owing to the absence of any form of magnifying device, its range is limited to about 30 m, its accuracy at that distance being + or −5 mm. It will be seen from these facts that it in no way supplants the ordinary form of level, but it has a considerable number of uses in the levelling of building works.

The optical system is formed of five mirrors A, B, C, D and E, mirror C being mounted on a pivoted and weighted arm so that no matter in what position the instrument is set the mirror will remain in a horizontal plane. The remaining mirrors are fixed. The system is in two

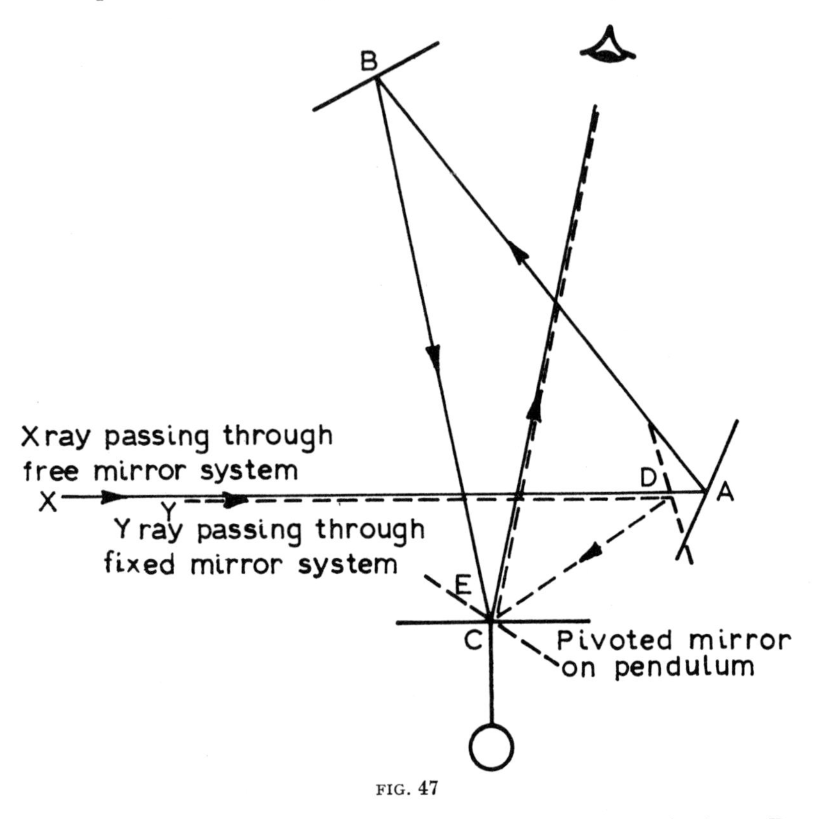

FIG. 47

parts—one consists of mirrors A, B and C, and the other of mirrors D and E. The two systems are fixed side by side (see fig. 47).

The staff is a short wooden one with a sliding crosspiece which can be fixed by means of a locknut (see fig. 48). The **back** of the staff is divided into metres and cm, and thus **the staff holder** records the levels taken.

USE—The instrument is set up on its tripod, no attempt being made to set it precisely level.

Two rays of light, X and Y (see fig. 47), project from the two ends of the crosspiece, ray X being reflected through system A, B and C, and ray Y through system D and E. Unless, fortuitously, the instrument has been set up exactly level, the image seen will be split as A in fig. 48.

To make the line of collimation level, the target is moved up or down until the images of the crosspiece come into conjunction as B (fig. 49). The level reading is then taken from the back of the staff.

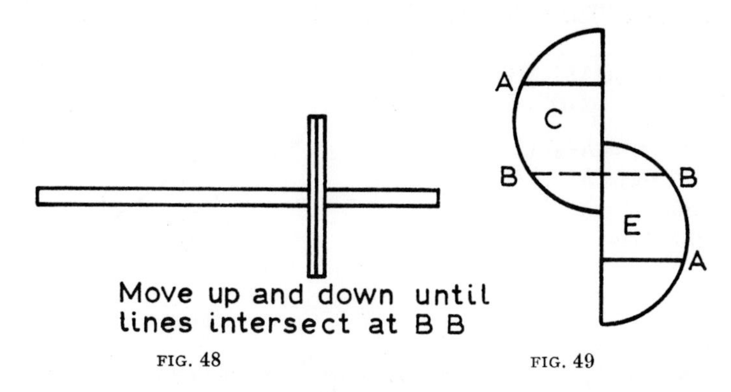

FIG. 48 FIG. 49

CURVATURE AND REFRACTION

CURVATURE—As the earth is curved, a horizontal sight does not give the true height above ground of any point. The error due to curvature is negligible for short sights, but adjustment is necessary for long sights.

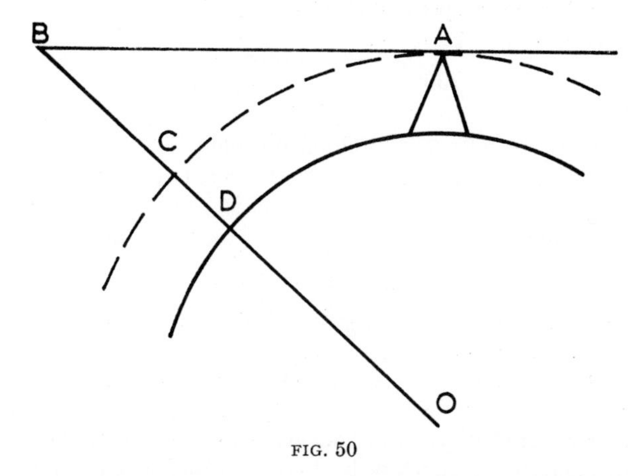

FIG. 50

Observing staff BD from A (see fig. 50), a difference BC results, and reduced level D is made lower by BC than it actually is.

Let AD in fig. 51 represent the earth's surface and O the centre of the earth. AB is the horizontal sight at A, i.e. at right angles to R, the earth's radius at A. BD is the curvature correction for distance AB, which is taken as equal to AD; the actual difference being negligible compared to the earth's radius.

To determine the curvature correction BD, i.e. c.

Now $OB^2 = OA^2 + AB^2$ and since OA = OD = R

$$(R+c)^2 = R^2 + X^2$$
$$R^2 + 2Rc + c^2 = R^2 + X^2$$
$$2Rc + c^2 - X^2$$
$$c(2R+c) = X^2$$
$$c = \frac{X^2}{2R+c}.$$

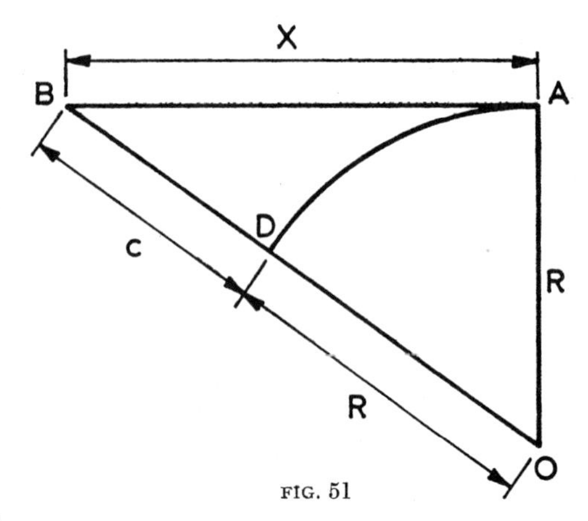

FIG. 51

Since c compared to 2R is extremely small, it may be ignored and the expression may be rewritten as:

$$c = \frac{X^2}{2R}$$

$$\text{i.e. Curvature correction} = \frac{\text{Distance}^2}{\text{Diameter of earth}} = \frac{d^2}{2R}$$

The curvature correction for a sight of 1 km length

$$= \frac{1}{12\,800} \times 1000 \times 1000 = 80 \text{ mm approx.}$$

REFRACTION

Rays of light are refracted by differing conditions of the atmosphere. This refraction bends the ray towards the earth. The magnitude of the correction to be applied due to refraction is $\frac{1}{7}$ of the curvature correction, and in the opposite direction. Taken together the correction:

$$= \frac{d^2}{2R}\left(1 - \frac{1}{7}\right)$$
$$= \frac{6}{7} \times \frac{d^2}{2R}.$$

RECIPROCAL LEVELLING

This is a method of levelling accurately across wide streams, etc., where equalisation of length of back and foresights is not possible.

The principle is that two sets of possibly erroneous differences are averaged to give the true differences. This method eliminates any errors due to curvature or refraction.

With level at C (see fig. 52) and staffs at A and B, read levels at A and B. Move level to D so that DB = CA, and read A and B again. Then the mean of the two differences of level is the correct difference.

Example *Difference*

Reading with instrument at C, A = 4, B = 2·5	1·5
Reading with instrument at D, B = 5, A = 6·1	1·1
	2)2·6
Correct difference =	1·3

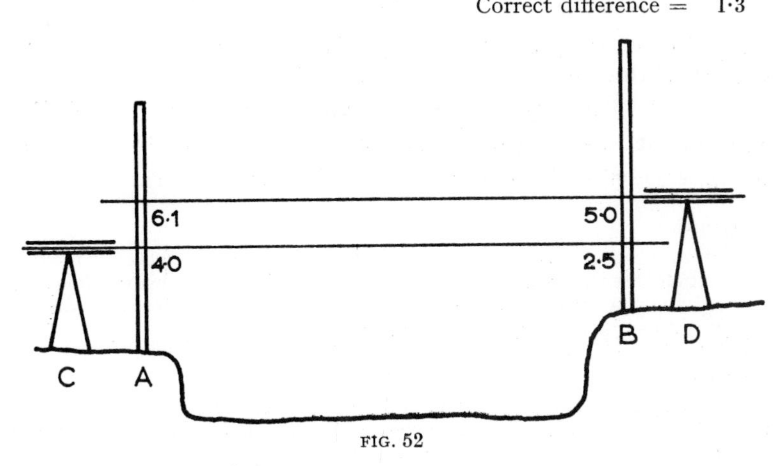

FIG. 52

SOURCES OF ERROR

INCORRECT SETTING-UP

1. Bubble off centre when the reading was taken. Check bubble before booking.
2. Movement of staff during the change-over of level.
3. Staff not held vertically. Use a circular bubble or plumb-bob.
4. Parallax. Adjust as previously described.
5. Instrument moved during readings of backsight and foresight.
6. Staff not properly extended.

INSTRUMENT ERROR AND CORRECTION

1. Collimation error. Check frequently and equalise length of back- and foresights.
2. Errors in staff graduation. Check before using.
3. Loose tripod head. Repair.
4. Telescope not parallel to bubble tube. Permanent adjustment.
5. Telescope not at right angles to the vertical axis. Permanent adjustment.

WEATHER CONDITIONS

Poor reading due to climatic conditions, i.e. heat haze, bad light, etc. Reduce length of sights.

Note—The maximum permissible length of sight with an ordinary instrument under reasonable conditions is 100 m.

MISTAKES

1. Incorrect readings taken.
2. Booking errors.

PERMISSIBLE ERROR

In work on building sites the permissible error is $\pm 0{\cdot}006$ on the closing bench mark.

When levelling long distances, the permissible error in metres $= \text{Constant} \times \sqrt{\text{Distance levelled in kilometres}}$, where the constant on moderate ground $= 0{\cdot}02$, and on rough ground $= 0{\cdot}03$.

Example

For what purpose is a dumpy level used? Describe a method of making the bubble 'traverse' so that it remains central while the instrument is rotated about a vertical axis. Also describe a method of testing and adjusting, if necessary, the horizontal line of collimation.

Example

The table shows the result of a levelling operation to determine the reduced level of a roof. Complete the table and apply the checks. One of the readings is taken on an inverted staff.

Backsight	*Inter. Sight*	*Foresight*	*Reduced Level*	*Remarks*
2·88			457·32	'C' corridor.
4·50		0·54		Step, first flight C–D.
	2·86			Landing between C and D.
4·48		0·53		Step, second flight C–D.
	2·85			'D' corridor.
4·55		0·54		Step, first flight to 'E'.
	2·90			Landing between D and E.
4·50		0·55		Step, second flight to E.
	2·88			'E' corridor.
3·02		−4·89		Floor-level roof shed.
	2·77			Roof outside shed.
2·98		2·84		S.E. corner roof.
	2·89			Centre S. end roof.
	2·99			S.W. corner roof.
	0·35			Parapet corner roof.
	2·94			N.W. corner roof.
	0·27			Parapet corner roof.
	2·89			Centre, N. roof.
	2·98			N.E. corner roof.
		1·65		Parapet corner roof.

Answer: Total Rise 25·15 m.

Solution

Backsight	*Inter. Sight*	*Foresight*	*Height of Inst.*	*Reduced Level*
2·88			460·20	457·32
		0·54		459·66
4·50			464·16	
	2·86			461·30
		0·53		463·63
4·48			468·11	
	2·85			465·26
		0·54		467·57
4·55			472·12	
	2·90			469·22
		0·55		471·57
4·50			476·07	
	2·88			473·19
		−4·89		480·96
3·02			483·98	
	2·77			481·21
		2·84		481·14
2·98			484·12	
	2·89			481·23
	2·99			481·13
	0·35			483·77
	2·94			481·18
	0·27			483·85
	2·89			481·23
	2·98			481·14
		1·65		482·47
26·91		6·65		482·47
1·76		4·89		457·32
25·15		1·76		25·15

Example

Show how curvature of the earth and atmospheric refraction can be eliminated when levelling over long distances. Find the correction to be applied to the apparent difference in levels between the stations $1\frac{1}{2}$ km apart, if the radius of the earth is taken as 6390 km and the level is set up near one of the stations. The correction for refraction may be taken as one-seventh that for curvature.

Answer: 0·150 m to be subtracted from reading on distant staff.

Solution

$$\text{Correction} = \frac{6}{7} \times \frac{d^2}{2R}$$

$$= \frac{6}{7} \times \frac{1{\cdot}5^2}{2 \times 6\,390} \times 1000 \times 1000$$

$$= 150 \text{ mm.}$$

Example

What is meant by the term 'reciprocal levelling'? Explain how this method used for long sights eliminates errors due to curvature or collimation adjustment.

Two pegs A and B separated by a river approximately 300 m wide, the reduced level of A being 50·382 m.

The level was set up at A, its eyepiece being 1·531 m above the peg. A sight on to a target staff at B read 2·038 m. When the instrument was transferred to B, the height of the eyepiece was 1·486 m above the peg, and the reading on the target staff at A was 2·012 m.

Find the reduced level of the peg at B.

Solution

	Difference
Readings with instrument at A, A = 1·531 m, B = 2·038 m .	0·507 m
Readings with instrument at B, A = 2·012 m, B = 1·486 m .	0·526 m
	2)1·033
Correct difference =	0·516 m

Reduced level of B = 50·382 − 0·516
= 49·866 m

Example

How does the curvature of the earth and terrestrial refraction affect surveying operations? A combined correction of $66{\cdot}9d^2$ is usually used in this connection, where d is the distance in kilometres and the correction is in millimetres. Show how this correction is derived.

Note.—In the script the formula is given as

$$\frac{6}{7} \times \frac{d^2}{2R}\text{; this equals } \frac{6 \times 1000 \times 1000}{7 \times 12\,800} d^2 = 66{\cdot}9d^2.$$

Example

The figures tabulated below are all that remain visible on a much battered page from a levelling book. By inspection of the remaining figures, fill in the queried entries. Tabulate the completed results by the 'Height of Instrument' method, but leave a space after chainage 5 + 00, and assume that it was necessary to turn over the page in the levelling book at this juncture. Below this space continue the booking as though you were starting at the top of a fresh page.

Also book the results by the 'Rise and Fall' method with a change of pages at the same chainage; in both methods make any necessary checks.

Chainage	*B.S.*	*Int.*	*F.S.*	*H.I.*	*R.L.*
1 + 00	2·99			222·26	?
2 + 00		3·24			219·02
3 + 00		?			218·80
4 + 00	?		4·18	221·47	218·08
5 + 00		3·23			218·24
6 + 00	4·01		3·51	?	217·96
7 + 00		3·22			218·75
8 + 00			2·48		?

Answer

Chainage	*B.S.*	*Int.*	*F.S.*	*Rise*	*Fall*	*R.L.*
1 + 00	2·99					219·27
2 + 00		3·24			0·25	219·02
3 + 00		3·46			0·22	218·80
4 + 00	3·39		4·18		0·72	218·08
5 + 00			3·23	0·16		218·24
	6·38		7·41 1·03	0·16	1·19 1·03	1·03

Chainage	*B.S.*	*Int.*	*F.S.*	*Rise*	*Fall*	*R.L.*
5 + 00	3·23					218·24
6 + 00	4·01		3·51		0·28	217·96
7 + 00		3·22		0·79		218·75
8 + 00			2·48	0·74		219·49
	7·24 1·25		5·99	1·53 1·25	0·28	1·25

Note—This problem is unlikely to arise in practice. If some of the figures were obliterated, it is likely that most of the figures would be partially obscured and it would be wiser to carry out a new survey if possible.

3 APPLICATIONS OF LEVELLING

THE applications of levelling are as follows:

(*a*) Determining the relative vertical heights between two stations (this has been dealt with under 'Method').

(*b*) Contouring.

(*c*) Planning of road, rail and similar services.

(*d*) Planning of underground services, such as drains, etc.

(*e*) Fixing site levels.

CONTOURING

DEFINITION OF A CONTOUR—A line drawn through all the points on the surface of the land which are at the same height; the line therefore divides the land which is higher than the given height from that which is lower.

Thus, if the sea is assumed to be at level 0·00, the shore-line would coincide with the contour of 0·00. If the sea were raised 100 m, the new shore-line would coincide with the contour of 100·000.

Fig. 53 shows the contour lines of a piece of land.

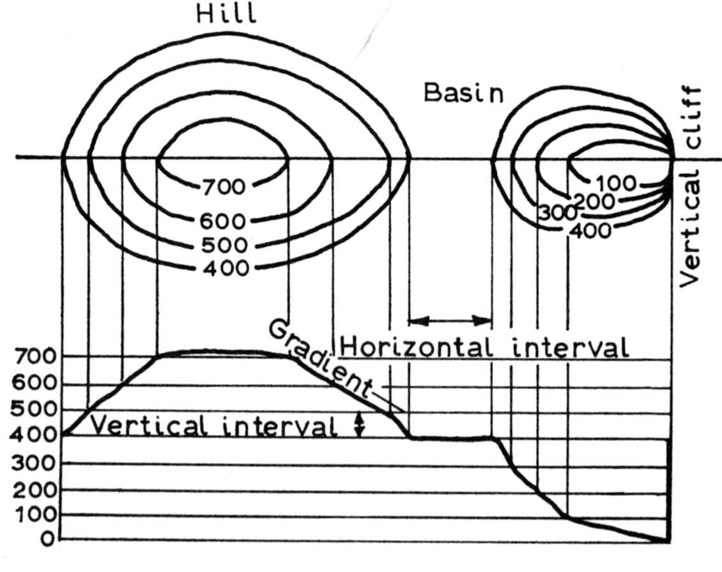

FIG. 53

VERTICAL OR CONTOUR INTERVAL is the difference in height between one contour line and the next.

The magnitude of the contour interval depends upon:

(a) the purpose of the map—the following are suggested intervals:
  (i) small building sites—1 m.
  (ii) large industrial sites—2–4 m.
  (iii) location surveys for services—2–10 m.

(b) the time and money available—the smaller the interval, the greater the amount of time and money required;

(c) the nature of the country—in a flat valley tract the interval should be small to depict the nature of the country. In mountain areas the contour intervals need not be small;

(d) the scale of the map—the greater the scale the smaller the contour interval.

HORIZONTAL EQUIVALENT is the distance between one contour line and the next measured on the plan.

$$\text{SLOPE or GRADIENT} = \frac{\text{Contour interval}}{\text{Horizontal equivalent}}.$$

The slope or gradient is usually expressed as a ratio of the height risen in a horizontal distance, i.e. 1 in 10 means a rise of 1 unit in a horizontal distance of 10 units.

The slope is sometimes expressed as a percentage, i.e. a 10 per cent. slope means a rise of 10 units in a horizontal distance of 100 units.

PROPERTIES OF CONTOURS

(a) Where the horizontal distance between contour lines is least, the gradient is greatest.

(b) Contour lines cannot cross.

(c) Contour lines must form continuous lines.

(d) Only in the case of vertical cliffs can contour lines join.

METHOD OF CONTOURING

(1) DIRECT METHOD

The level is set up in a field to be contoured and its height of collimation determined. Suppose the backsight is 2·350 on to a B.M. of 95·000, then the height of collimation is 97·350. If the staff is held on the ground in the field to be contoured and a reading of 1·350 is obtained, then the ground at the staff station is at reduced level of 96·000. If the reading on the staff is not 1·350, the staff is moved about the field until the reading is 1·350. When the required reading is obtained, a peg is inserted at the staff position. This process is repeated until a sufficient number of pegs are placed to locate the contour line of 96·00. If the next contour line is to be 94·000, then the staff is moved until a staff reading of 3·350 is obtained. It is advisable to mark on the peg the reduced level of the ground or to paint bands of colour on the pegs, one colour representing one contour throughout and each contour having a different colour (see fig. 54).

After all the contour lines have been pegged, the pegs are located by offsets from the existing chain lines (see fig. 54).

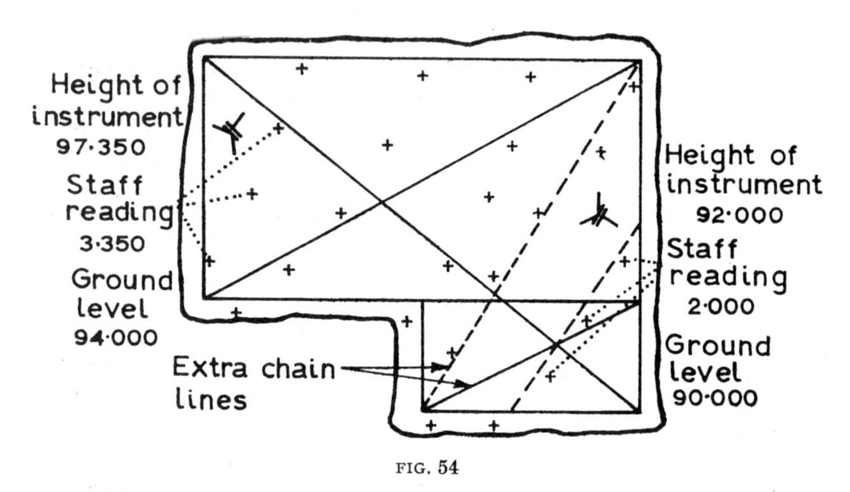

FIG. 54

Advantages of Method

(*a*) Most accurate method.
(*b*) Requires little office work in plotting contours.

Disadvantages

(*a*) Slow and therefore costly in the field.
(*b*) Extra chaining has to be carried out to locate the pegs.
(*c*) Suitable only for small, fairly regularly sloping sites requiring small contour interval.

(2) GRID METHOD

The grid of lines is set out on the site (see fig. 55) (the longest chain line is often used as a base).

The distance between the grid lines depends upon the nature of the slope. If the slope is gentle, the grid lines can be put in at 30 m apart. The ground level is then taken at each line intersection. The plan of the site and the grid lines is plotted. The position of the contour lines is found by interpolation or by using an interpolation graph.

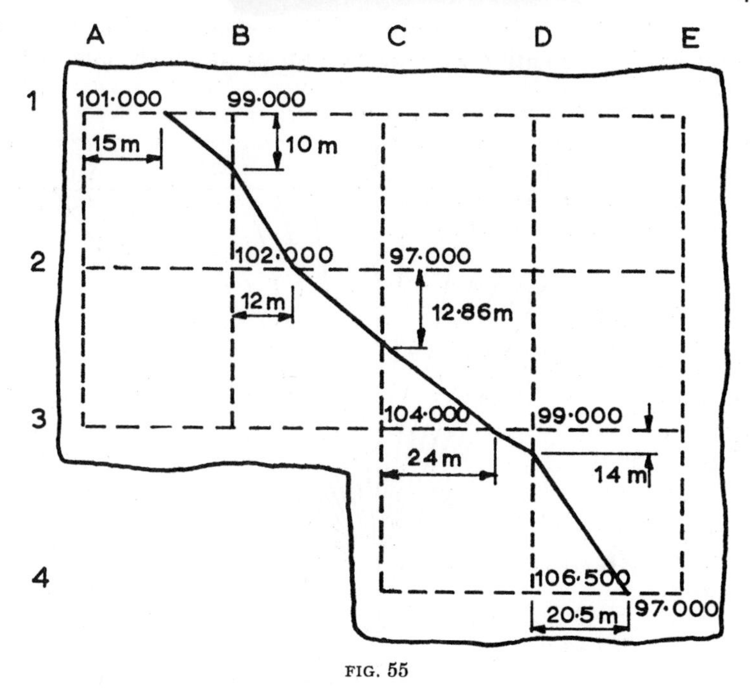

FIG. 55

Example

It is required to plot the contour line of 100·00 shown in fig. 55. The calculation sheet is given below:

Stations	*Fall or Rise*	*Contour Fall or Rise*	*Distance*	*From*	*To*
A1–B1	101·000 − 99·000 = 2	101 − 100 = 1	$\frac{1}{2} \times 30 = 15$ m	A1	B1
B1–B2	99·000 − 102·000 = 3	99 − 100 − 1	$\frac{1}{3} \times 30 = 10$ m	B1	B2
B2–C2	102·000 − 97·000 = 5	102 − 100 − 2	$\frac{2}{5} \times 30 = 12$ m	B2	C2
C2–C3	97·000 − 104·000 = 7	97 − 100 = 3	$\frac{3}{7} \times 30 = 12{\cdot}86$ m	C2	C3
D3–D4	99·000 − 106·500 = 7·5	99 − 100 = 1	$\frac{1}{7{\cdot}5} \times 30 = 4$ m	D3	D4
D4–E4	106·500 − 97·000 = 9·5	106·5 − 100 = 6·5	$\frac{6{\cdot}5}{9{\cdot}5} \times 30 = 20{\cdot}5$ m		

INTERPOLATION GRAPH (see fig. 56)

A line, XX, is drawn on a sheet of drawing-paper and is marked off into a number of equal parts (any multiple of ten is convenient), and each mark is numbered. From the centre of the line at Y a perpendicular YY′ is erected. Rays are then drawn from Y′ to the marks set off along XX. It is convenient to draw every tenth line in a different coloured ink. The graph is placed under a plan drawn on tracing-paper, and moved until the line marked 101 on the graph is under the grid point of level 101.

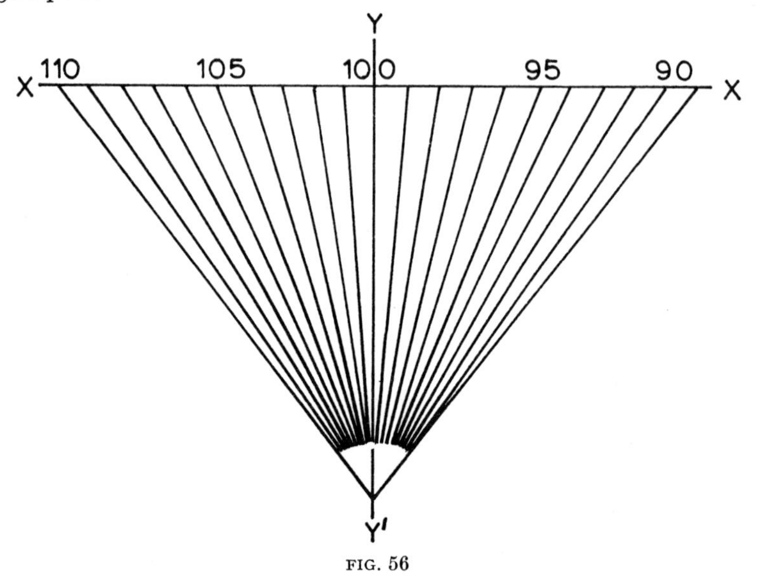

FIG. 56

The graph is moved backwards or forwards until the grid points A′, level 110, is over the grid line 110 and grid point B′, level 90, is over grid line 90. Then the point on the grid line A′, B′, which is over the

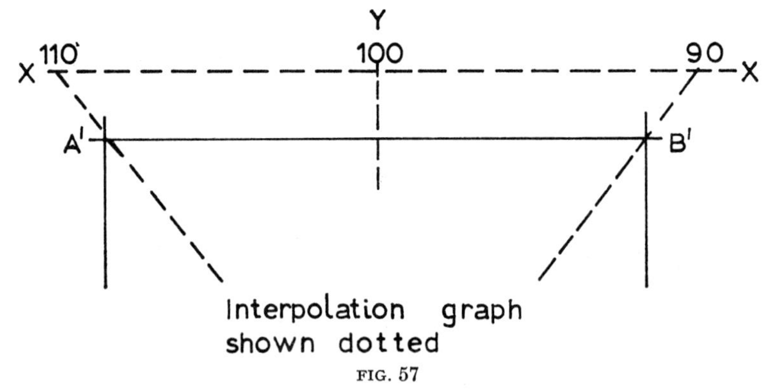

FIG. 57

graph lines of 100, is at the contour line of 100, and this point is marked on the plan (see fig. 57). (The student is advised to make an interpolation graph, and to carry out the above process, which may seem complicated but is simple in practice.) Normally the grid levels will not be an exact number of metres, and allowance must be made for the fractions of a metre.

Advantage of the Grid Method

The average level of the site can easily be determined (see below).

Disadvantages

(*a*) It is based on the assumption that the ground slopes uniformly between grid stations and ignores changes of level inside the squares.

(*b*) It takes a long time on site.

(3) TACHEOMETRIC METHOD (see page 141)

The position of the levels are chosen at change of slope and the position and level found by tacheometry.

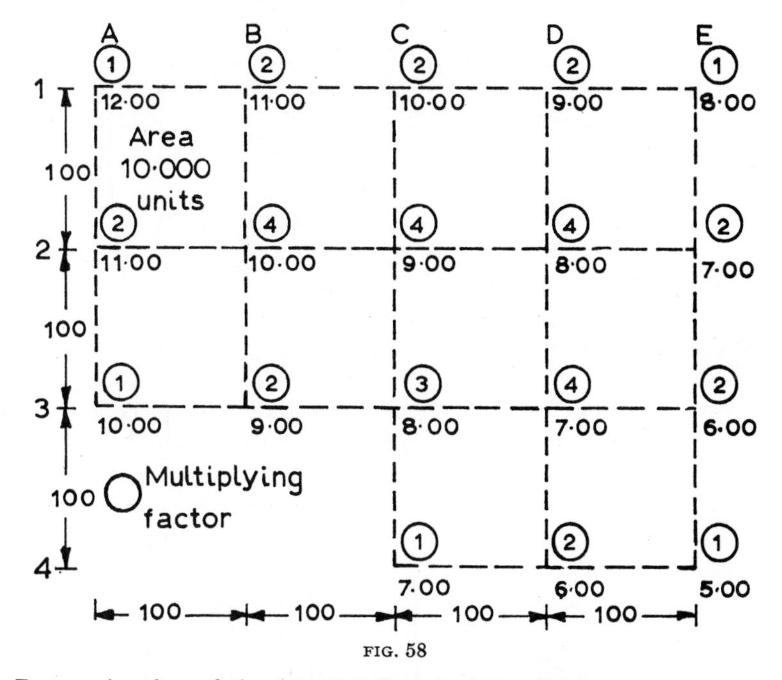

FIG. 58

Determination of the Average Level of the Grid

The reduced level at each grid point is multiplied by a factor, depending on the point's position. The sum of the multiplications is then divided by four times the number of squares.

The factor depends upon the number of squares meeting at the grid point, e.g. at grid point A1 (see fig. 58) there is only one square, therefore the reduced level at this point is multiplied by one. At B1 there are two squares meeting at the point, therefore the reduced level at B1 is multiplied by 2. At B2 the reduced level is multiplied by 4, and at C3 the reduced level is multiplied by 3.

USE OF CONTOUR MAPS

(*a*) **Location of possible routes** for roads, drains, etc. In fig. 59, A and B are the terminals of a proposed road. If the road was run in a straight line from A to B, it would have to descend into the valley from A, cross one river, climb the hill, descend into the next valley,

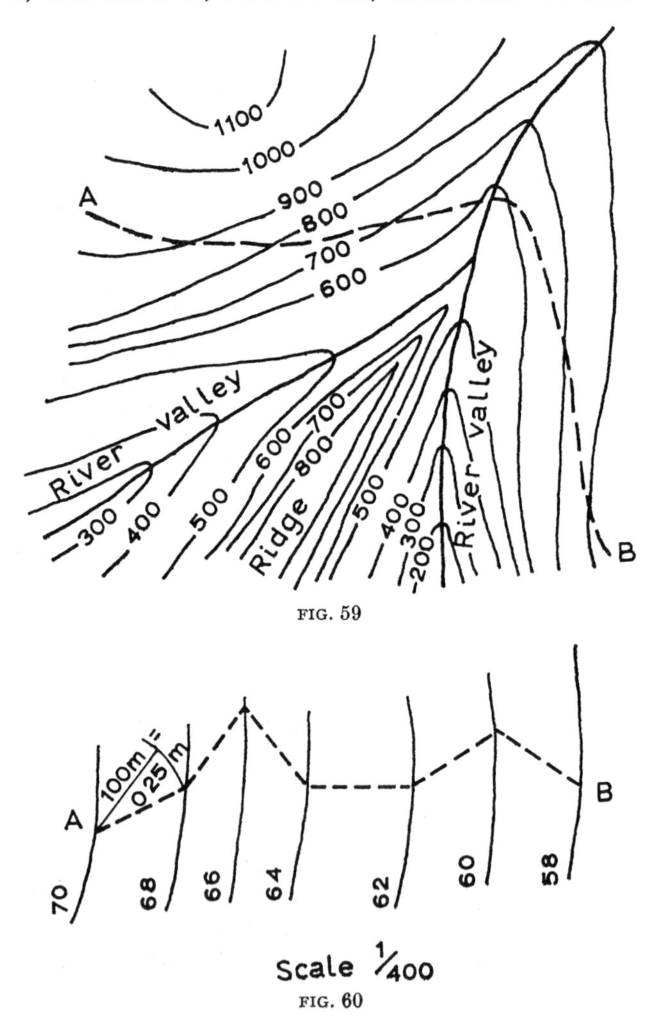

FIG. 59

FIG. 60

cross over the second river and climb again. A possible solution to give reasonable gradients is shown by the dotted line.

In fig. 60, A and B are the inlet and outlet respectively of a sewer which is to have a fall of 1 in 50, therefore when falling from one contour line to another a fall of 2 m, the drain must travel a horizontal distance of 100 m. If the scale is 1 in 400, the dividers are opened to 0·25 m, one leg pinned at A and an arc struck where the leg meets the 68 m contour. The dividers are then swung from this point, and another point is marked where the leg meets the 66 m contour. The intersection points are joined up, giving a possible route for the sewer.

(*b*) **Laying out Building Sites.**—The position of hilltops, basins, steep slopes, etc., can be seen from contour plans. Buildings should not generally be sited on exposed hilltops, steep slopes requiring stepped foundations and risking possible soil creep, or in basins which may form natural drainage areas.

(*c*) **Calculation of Volumes**—Fig. 61 shows the site of a proposed reservoir. To determine the volume of water impounded, the area of

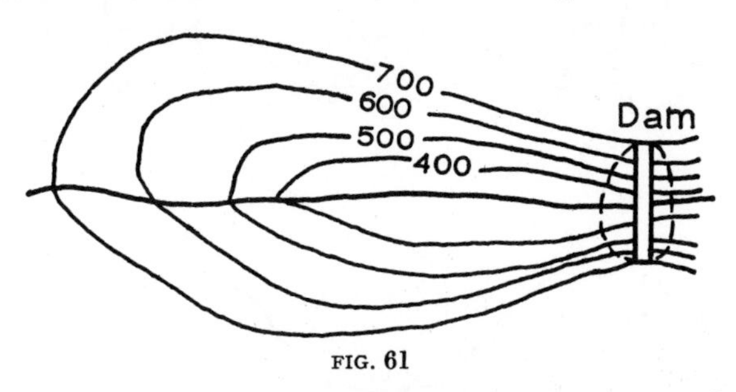

FIG. 61

land encircled by each contour and the dam is measured and the volume calculated by using Simpson's Rule or the Trapezoidal Rule (see pages 70 & 71).

(*d*) **Determination of Intervisibility.**—When planning stations for tunnel surveys, plane-table surveys, etc., the stations must be intervisible. The stations are marked on the plan, and sections drawn along the line to see if they are intervisible. In fig. 62 it can be seen that stations X and Y are not visible from station C, and extra stations, A and B, are required.

PLANNING OF ROADS

The planning of main roads is a specialised and highly skilled occupation, and it is not proposed to cover such works in this book. It is intended to deal with estate roads and similar small works.

The requirements of such works fall into two main groups: alignment, and level.

ALIGNMENT

(*a*) Straightness. The straighter the road, generally the less materials required for construction, the faster the movement of traffic. Where changes of direction are imperative, the change should be made on curves of large radius giving good visibility.

LEVEL REQUIREMENTS

(*a*) Should run over ground of uniform, gentle slopes.

(*b*) Should run over ground not exceeding maximum gradients.

(*c*) 'Cut to Balance Fill' (for detailed information see page 76). Where route is aligned so that cuttings and embankments are necessary, the road level should be fixed so that the volume of earth excavated is equal to the volume of earth required for embankments.

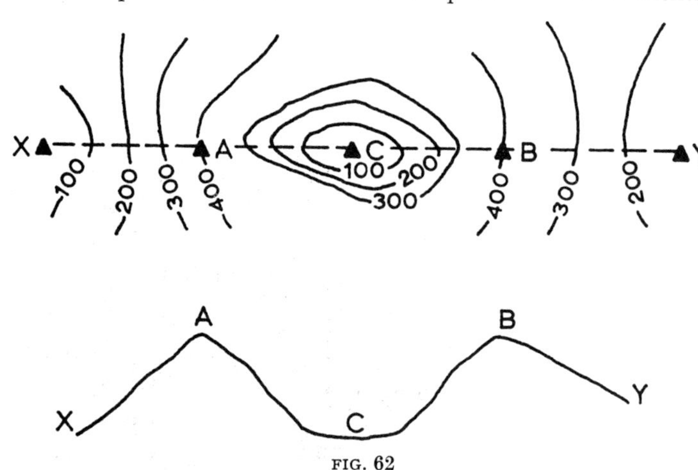

FIG. 62

CHOOSING ROUTE

Cases may arise where the desired requirements are incompatible, the procedure to be adopted in such cases is outlined below.

(*a*) **Reconnaissance**, either on the site or from contoured maps and three or four alternative routes chosen.

(*b*) **Rough traversing and sectioning** is carried out using such instruments as the **prismatic compass** and **Abney level** (see Chapter 5).

(*c*) **Strip maps and sections** are then drawn for each alternative route, and one is decided upon.

(*d*) **Accurately survey the decided route.**

Method of Surveying Decided Route

(*a*) **Set out the centre line** with the **theodolite** (see Chapter 5). The pegs should be at 30 m centres on uniformly sloping ground and at 10 m on rough ground (see fig. 63).

(*b*) **Set out grid lines** parallel to the centre line at ½, and 1½ times the road width on either side of the centre line (where deep cuttings or high embankments are likely to be constructed, further lines, 2½ times the road width, should be set out).

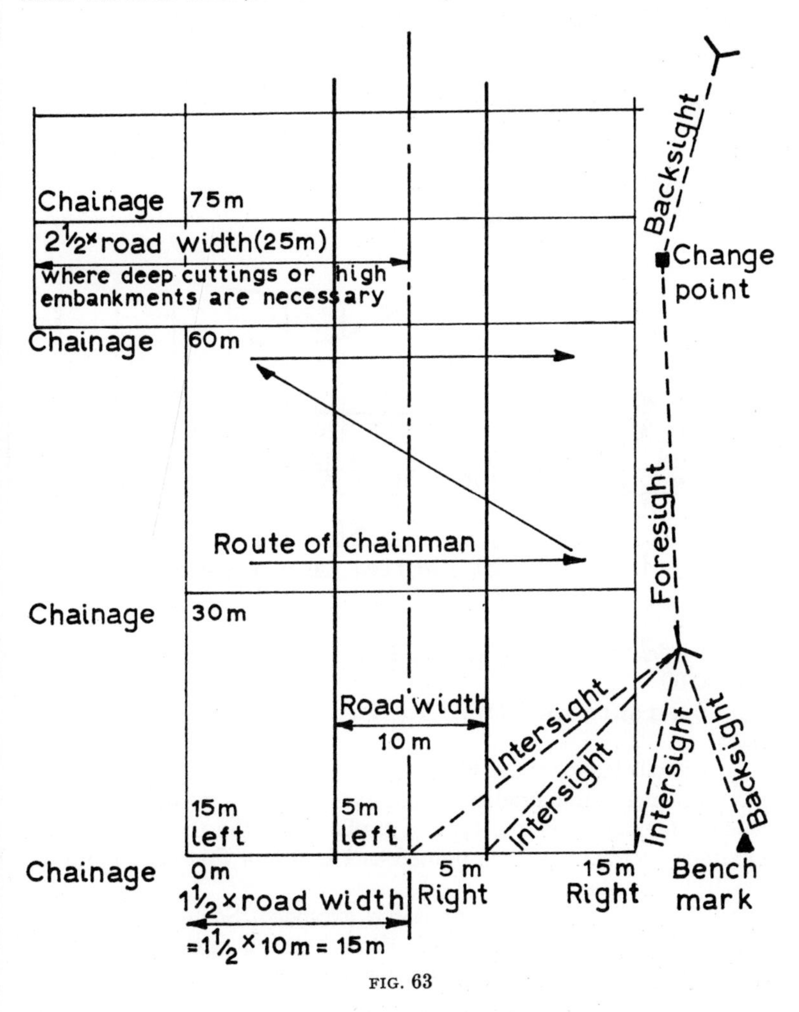

FIG. 63

(*c*) **Take levels at the grid points**—Where chainage cuts the set-out lines the chainman should traverse the grid from left to right consistently and not go from left to right on one cross line, then from right to left on the next; mistakes are likely to occur in the booking of the levels if this is allowed.

OFFICE WORK

(*a*) **Plot longitudinal section** of ground level along centre line of road (see fig. 64). The scales should be in the ratio of 1 to 10.

The scales commonly used are 1 in 1000 horizontally and 1 in 100 vertically. The height of the datum line should be fixed so that there is a 100 mm space between it and the lowest ground level. To do this, find the lowest ground level in the field book, deduct 10 m from it (10 m = 100 mm on the section), and bring the resultant figure to the nearest multiple of 1 m and make the datum at that level, i.e. the lowest

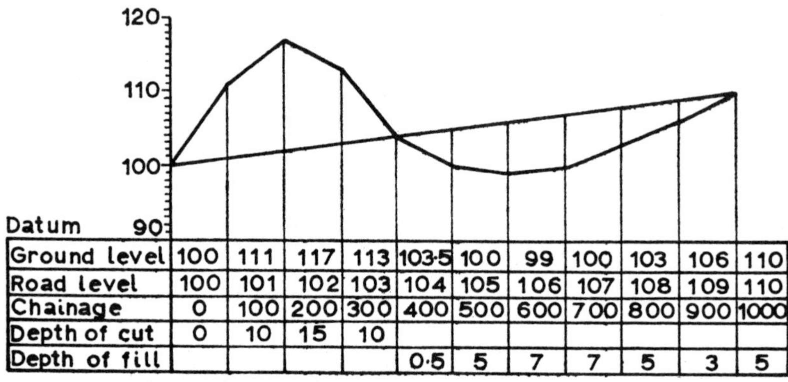

Ground level	100	111	117	113	103·5	100	99	100	103	106	110
Road level	100	101	102	103	104	105	106	107	108	109	110
Chainage	0	100	200	300	400	500	600	700	800	900	1000
Depth of cut	0	10	15	10							
Depth of fill					0·5	5	7	7	5	3	5

Longitudinal section along centre line

FIG. 64

ground level is 100·00 m, deduct 10 m, giving 90·00, making datum level 90·00.

(*b*) **Determine road level:** The final road level must be fixed so that:

(i) The cut balances the fill.

(ii) The maximum gradient is not exceeded. The maximum gradient depends mainly upon the type of traffic using the road surface. The maximum gradient is often quoted as 1 in 30 for first-class roads and 1 in 15 for estate roads.

(iii) The minimum gradient is not exceeded. Some gradient is required to ensure rain-water run-off and its magnitude will depend upon the type of surface finish; for example, on concrete roads it should be about 1 in 100.

(iv) Changes of gradient to be connected by vertical curves. A commonly used form of vertical curve is the parabola. The curve should give a clear view of about 100 yards.

(Calculations for vertical curves are given at the end of the chapter.)

(*c*) **Draw cross-sections**—The cross-sections should be drawn at regular intervals along the line of the road. The vertical and horizontal scales should be the same (this facilitates the area measurement of the cut or fill for volume calculations). The remarks about the fixing

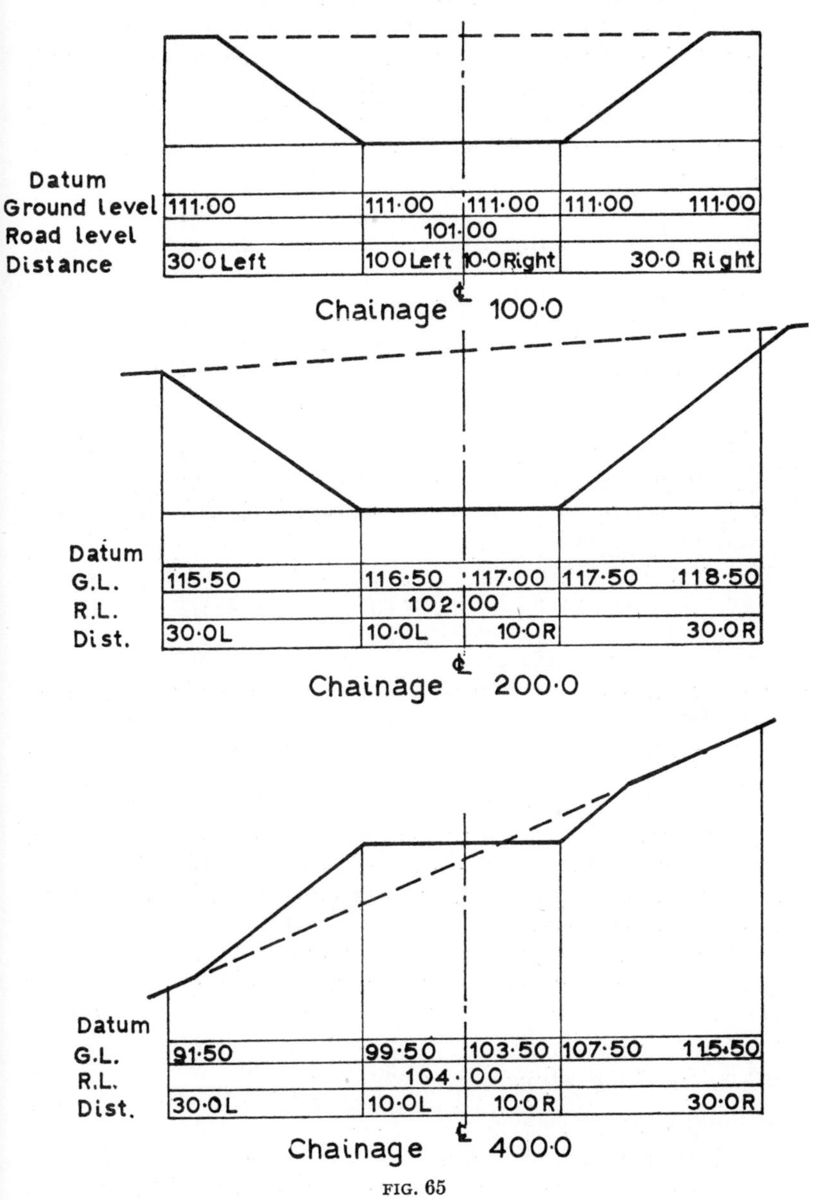

FIG. 65

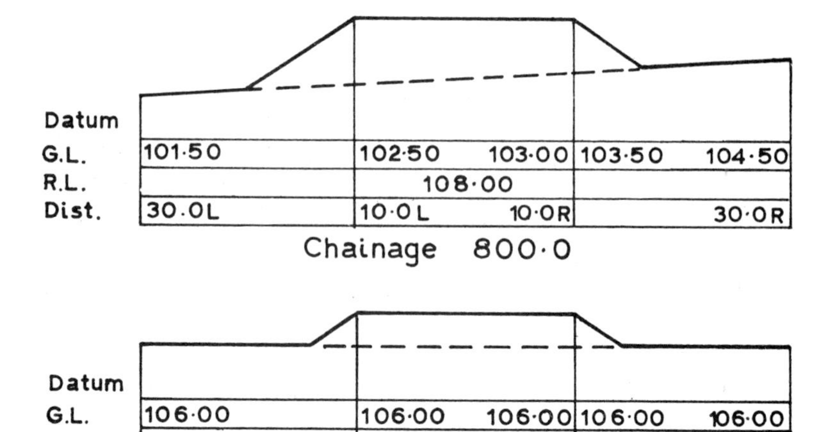

FIG. 66

of the datum line for the longitudinal section apply to cross-sections with the addition that the datum should be 10 m below road level when the road level is lower than the ground level, as in cuttings. The slope of the sides depends upon the material, its drainage, climatic conditions, etc. A common slope is 1 in $1\frac{1}{2}$ (see figs 65 and 66).

(*d*) **Determine earthwork volumes.**

(i) Determine the area of earthwork at each cross-section.

(ii) Apply Simpson's rule and determine the volume (see page 70).

(iii) Draw mass-haul curve (see page 76).

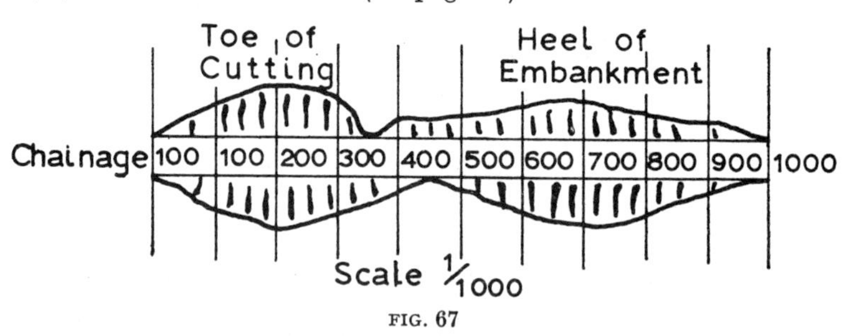

FIG. 67

(*e*) **Draw plan**—Draw a plan showing the heel of the embankment and the toe of the cutting (see fig. 67).

CONSTRUCTION LEVELLING

(*a*) Set slope stakes.

(*b*) Erect sight rails.

Setting Slope Stakes

A slope stake is a peg set in the ground at the intersection of the ground and the slope of the cutting or embankment. If a plan of the road has been drawn, the distance from the centre line to the intersection can be scaled from the plan and measured out in the field. If a plan has not been drawn, then the slope stakes must be set by the following trial and error method (see fig. 68).

(i) Find, from the longitudinal section the road level at the required position, say 120·00.

(ii) Determine the height of instrument of the level, say 118·00, and calculate grade staff reading (this = road level − height of instrument, i.e. 120·00 − 118·00 = 2·00).

(iii) Find the reduced level of the ground at the centre line, say 107·00, and let the side slopes be 1 in $1\frac{1}{2}$.

(iv) Chainman holds the staff at a roughly estimated position of the toe of the slope and measures the distance from the staff to the centre line.

(v) The staff is read, say it is 1·80, this represents a fall of 1·80 + 2·00 = 3·80 m. If the staff is held at the exact intersection of

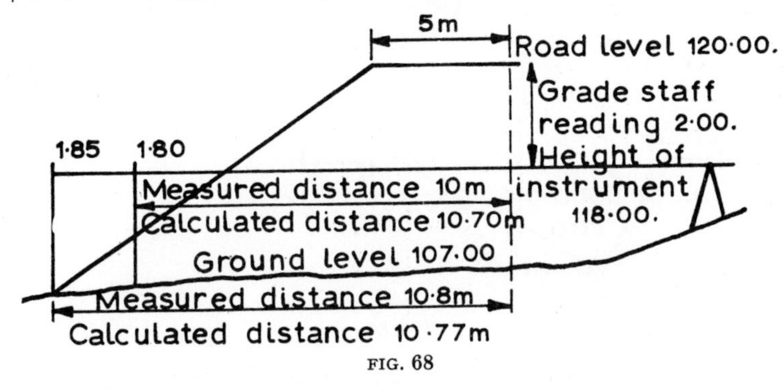

FIG. 68

slope and ground, its distance from the centre line should equal $3{\cdot}80 \times 1\frac{1}{2} + 5{\cdot}00$ ($\frac{1}{2}$ road width) = 10·70 m. If the staff is at 10 m from the centre line, it should be moved back to 10·70 m from the centre line.

(vi) The staff is reread, say it is now 1·85 the distance should be $(1{\cdot}85 + 2{\cdot}00) \times 1\frac{1}{2} + 5{\cdot}00 = 10{\cdot}77$ m. If a peg is now placed at 10·8 m from the centre line, it will be close enough to the intersection for all practical purposes.

Sight Rails for Cuttings

Sight rails consist of a board fixed horizontally to two stout pegs driven firmly into the ground (see fig. 69). Sight rails for each section must all be at the same height above the road level, and the height above the road should be some multiple of 100 mm. The boards should be painted alternate bands of colour such as red and white, and have their height above road level clearly marked on them. They should not be placed so low above the ground that the ganger has to lie in the mud to sight nor so high that he has to stand on a box. A convenient height is 0·80 to 1·0 m above the ground. Boning rods of length equal to the height of the sight rail above the road level + depth of formation (formation is the depth from road surface to the underside of the road base), are made of short pieces of wood nailed at right angles to a long piece of board. When the top of the boning rod coincides with the line of sight of the sight rails, the excavation is at the correct level. If the top of the boning rod projects above the line of sight of the sight rails, then the cutting is not deep enough.

When the cutting is at the correct depth, level pegs are placed in it to give levels for the drains and formation.

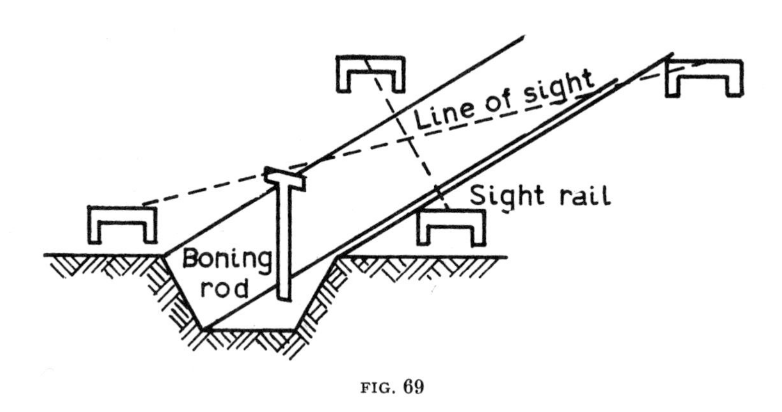

FIG. 69

PLANNING OF UNDERGROUND SERVICES—DRAINS

The choosing of the route follows the same procedure as outline for roads.

Office Work

(i) Plot longitudinal section.

(ii) Determine invert level (invert level, I.L., is the lowest part of the inside of the pipe) so that:

(*a*) The cover to the pipe is kept to a minimum. Cover is the depth of earth from the top of the pipe to the ground level.

(*b*) The maximum and minimum gradients are not exceeded.

Example

The following reduced levels (in metres) were obtained along the route of a drain:

Ground level .	104·46	106·13	108·25	107·61	108·62	108·82	108·94
Chainage . .	0	50	100	150	200	250	300
Ground level .	109·27	110·11	111·04	111·87	112·62	113·37	
Chainage . .	350	400	450	500	550	600	

Determine the invert levels of the drain, given the following data:
Maximum gradient 1 in 40. Minimum gradient 1 in 80. I.L. at 0 to be 100·00, minimum depth from ground to I.L. to be 3 m.

Solution

(i) Plot longitudinal section.

(ii) Fix outlet and peak I.L. The outlet is usually fixed, in this case it is 100·00, the peak is fixed to give minimum cover. The peak I.L. in this case may be **110·37**.

(iii) Fix I.L. along the line of the drain, remembering that the I.L. gradient should follow the ground level gradient as closely as possible to minimise excavation.

From 0 to 100 — The ground is at a steeper gradient than the maximum gradient of the drain. To minimise excavation, the drain can be placed at its maximum gradient.

From 350 to 600 — The ground has a gradient of approximately 1 in 60, and the drain can be put in at this gradient.

From 200 to 350 — The ground is at a shallower gradient than the minimum gradient of the drain. The drain will have to be placed at a gradient of 1 in 80.

From 100 to 200 — The drain gradient is 'fitted in' between the levels fixed above, the gradient being 1 in 55.

See over for Table

Construction Levelling

The position of the manholes is fixed. Level pegs are placed near them and profiles (see fig. 154) erected. The profiles must all be at the same height above I.L. If the trench is dug by machine, the profile posts must be wide enough apart to allow clearance for the machine, and the sight rail must not be fixed to the posts but supported by brackets. When the machine is close to the profile, the sight rail is lifted off the brackets and replaced when the machine has passed (see fig. 70).

When the trench has been excavated, level pegs should be driven into the trench to the height of the base concrete.

If the drain is to be water tested, construction should start from the outlet and not the peak, so that after the test the water can be disposed of easily. This method is advisable even when the water test is not used, for if heavy rain should occur the trench is liable to be flooded.

Chainage	0	50	100	150	200
Ground Level	104·46	106·13	108·25	107·61	108·62
Invert Level	100·00	101·25	102·50	103·41	104·32
Invert Level Gradient	1 in 40		1 in 55		

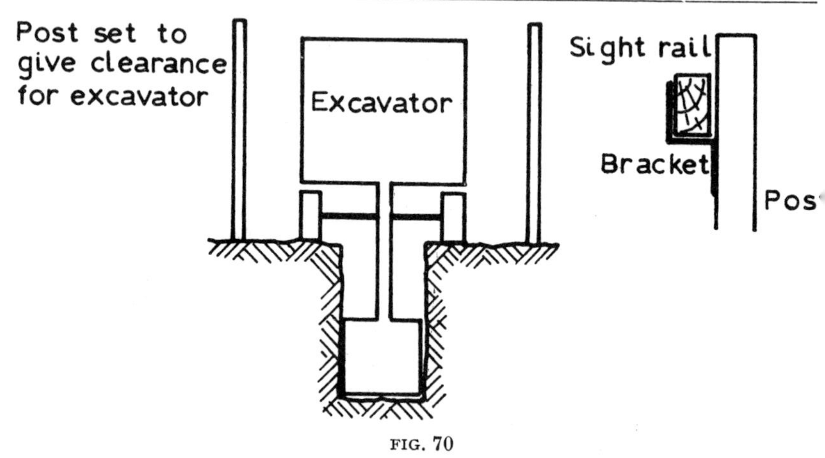

FIG. 70

CALCULATIONS FOR VERTICAL CURVES

Produce approach gradients to intersect at C.

1. Set back along gradients a distance of 100 m from intersection C, to A and B (see fig. 71).

2. Join A and B, bisect AB at D and join DC. The curve passes through E where EC = ED (this is a property of the parabola).

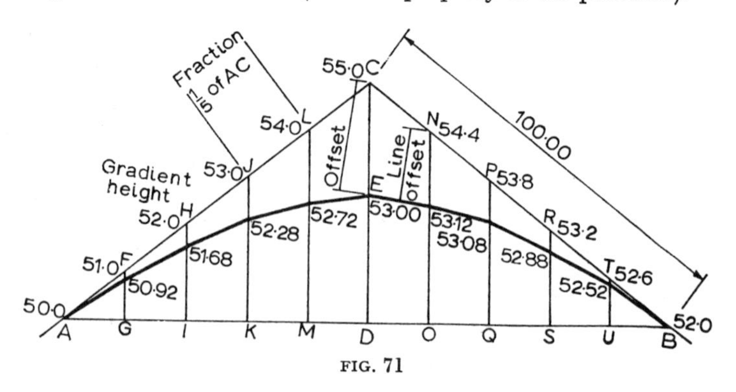

FIG. 71

250	300	350	400	450	500	550	600
108·82	108·94	109·27	110·11	111·04	111·87	112·62	113·37
104·94	105·57	106·20	107·03	107·87	108·7	109·53	110·47
1 in 80		1 in 60					

3. Divide AC and CB into a convenient number of equal parts, say five. Draw lines through these points parallel to ED to meet AB.
4. Calculate the heights of the points F, H, J, etc.:

$$\text{e.g. Height at F} = 50 + \tfrac{1}{5}(55 - 50) = 51.$$
$$\text{Height at H} = 50 + \tfrac{2}{5}(55 - 50) = 52.$$

5. Calculate the offsets along FG, HI, etc. (offsets from a tangent are proportional to the square of the distance along the tangent, this is another property of the parabola).

E.g. To find the height of road at line FG. The fraction of the distance along the tangent is $\frac{1}{5}$, this squared is $\frac{1}{25}$. This when multiplied by the length EC gives the offset down from the height F, i.e. $\frac{1}{25} \times 2{\cdot}0 = 0{\cdot}08$, $51 - 0{\cdot}08 = 50{\cdot}92$. The calculations are usually set out in tabular form as shown below.

It will be noted that lines CD are not vertical and that therefore the calculations of the road levels are inaccurate, the error is, however, negligible.

Line	*Fraction*	*Fraction² × Offset = Line Offset*	*Gradient Height − Line Offset = Road Level*
FG	$\frac{1}{5}$	$\frac{1}{25} \times 2{\cdot}0 = 0{\cdot}08$	$51{\cdot}0 - 0{\cdot}08 = 50{\cdot}92$
HI	$\frac{2}{5}$	$\frac{4}{25} \times 2{\cdot}0 = 0{\cdot}32$	$52{\cdot}0 - 0{\cdot}32 = 51{\cdot}68$
JK	$\frac{3}{5}$	$\frac{9}{25} \times 2{\cdot}0 = 0{\cdot}72$	$53{\cdot}0 - 0{\cdot}72 = 52{\cdot}28$
LM	$\frac{4}{5}$	$\frac{16}{25} \times 2{\cdot}0 = 1{\cdot}28$	$54{\cdot}0 - 1{\cdot}28 = 52{\cdot}72$
TU	$\frac{1}{5}$	$\frac{1}{25} \times 2{\cdot}0 = 0{\cdot}08$	$52{\cdot}6 - 0{\cdot}08 = 52{\cdot}52$
RS	$\frac{2}{5}$	$\frac{4}{25} \times 2{\cdot}0 = 0{\cdot}32$	$53{\cdot}2 - 0{\cdot}32 = 52{\cdot}88$
PQ	$\frac{3}{5}$	$\frac{9}{25} \times 2{\cdot}0 = 0{\cdot}72$	$53{\cdot}8 - 0{\cdot}72 = 53{\cdot}08$
NO	$\frac{4}{5}$	$\frac{16}{25} \times 2{\cdot}0 = 1{\cdot}28$	$54{\cdot}4 - 1{\cdot}28 = 53{\cdot}12$

Example

The following readings in metres were obtained in levelling a section for a proposed road, namely **4·323**, **3·373**, **1·154**, **4·186**, **1·729** and **4·486**. The fourth and fifth readings were on a change point off the line of section. The remainder were at points 100 m apart starting with the first reading at a point P, whose chainage is 1600 m and height above datum 39·823 m.

Make up the level book and if the formation level at P is to be **40·781** and the rise 1 in 50, find the depth of earthworks at each 100 m.

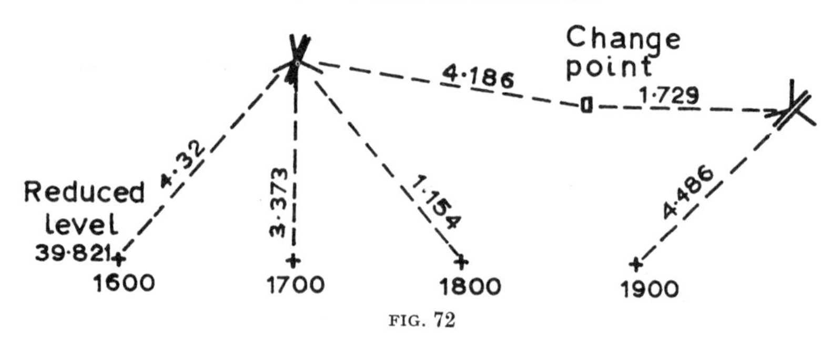

FIG. 72

Solution (see fig. 72)

Back-sight	*Inter. sight*	*Fore-sight*	*Height of Inst.*	*Re-duced Level*	*Chain-age*	*Form. Level*	*Cut*	*Fill*	*Remarks*
4·323			44·146	39·823	1600	40·781		0·958	Point P.
	3·373			40·773	1700	42·781		2·008	
	1·154			42·992	1800	44·781		1·789	
		4·186		39·960					
1·729			41·689						
		4·486		37·203	1,900	46·781		9·558	
6·052		8·672		39·823					
		6·052		37·203					
		2·620		2·620					

Note.

Depth of fill at 1600 = 40·781 − 39·823 = 0·958
Depth of fill at 1700 = 42·781 − 40·773 = 2·008
Depth of fill at 1800 = 44·781 − 42·992 = 1·789
Depth of fill at 1900 = 46·781 − 37·203 = 9·558.

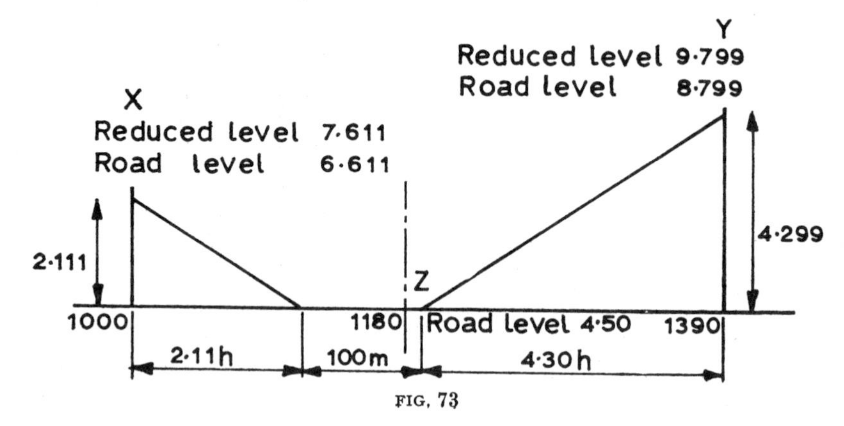

FIG. 73

Example

The field notes for levels along the centre line of a proposed road are as follows:

Backsight	*Intermediate*	*Foresight*	*Distance (metres)*	*Remarks*
0·542			1000	At point X.
	1·200		1030	
	3·362		1060	
1·877		4·848	1090	
	2·365		1120	
	4·306		1150	
4·744		4·402	1180	At point Z on bank of stream.
	3·414		1210	
	1·443		1240	
4·991		0·874	1270	
	4·867		1300	
	4·391		1330	
4·366		1·845	1360	
		2·363	1390	At point Y.

Make out the level book and reduce the levels, using as a datum the water level of the stream, which is 0·78 m below that of Z. There is to be a level portion of the road in the vicinity of Z, 100 m in length and 4·50 m above the datum. This is to connect two equal gradients commencing 1 m below ground level at X and Y, and falling towards Z. Find the gradient.

Solution (see fig. 73)

Back	*Int.*	*Fore*	*Rise*	*Fall*	*R.L.*	*Distance metres*	*Remarks*
0·542					7·611	1000	X
	1·200			0·658	6·953	1030	
	3·362			2·162	4·791	1060	
1·877		4·848		1·486	3·305	1090	
	2·365			0·488	2·817	1120	
	4·306			1·941	0·876	1150	
4·744		4·402		0·096	0·780	1180	Z
	3·414		1·330		2·110	1210	
	1·443		1·971		4·081	1240	
4·991		0·874	0·569		4·650	1270	
	4·867		0·124		4·774	1300	
	4·391		0·476		5·250	1330	
4·366		1·845	2·546		7·796	1360	
		2·363	2·003		9·799	1390	Y
16·520		14·332	9·019	6·831			
2·188			2·188		2·188		

Road level at X = 7·611 − 1·000 = 6·611.
Road level at Y = 9·799 − 1·000 = 8·799.
Fall from road level at X to level section = 6·611 − 4·500 = 2·111.
Fall from road level at Y to level section = 8·799 − 4·500 = 4·299.
Let gradient = 1 in h.

Then:

Horizontal distance from X to level = $2{\cdot}11h$.
Horizontal distance from Y to level = $4{\cdot}30h$.
Total horizontal distance from X to Y = 1390 − 1000 = 390 m.
Total horizontal distance from X to Y = $2{\cdot}11h + 100 + 4{\cdot}30h$.

$$\therefore \quad 390 = 2{\cdot}11h + 250 + 4{\cdot}30h.$$
$$140 = 6{\cdot}41h$$
$$h = \frac{140}{6{\cdot}41} = 21{\cdot}8$$

Gradient = 1 in 21·8.

Example

When making a preliminary assessment of the width of land required for a new road after selecting a trial grade line, a surveyor used a trial-and-error method by taking level readings at the centre line and at various distances along the cross-section.

At a particular cross section, the ground level at the centre line was 112·32 above datum, whilst the formation level was 115·00 m. Staff readings at the centre line, 12 m right and 15 m right from it, were respectively 1·88, 4·89 and 5·18. The formation width is 10 m, side slopes $1\frac{1}{2}$ horizontally to 1 vertical.

(A) Describe the basis of a method which will enable the toe of the embankment to be established by trial and error, giving any formula required.

(B) Show that this point is neither at 12 m nor at 15 m right from the centre line in the case considered.

(C) Assuming that the cross fall is constant, calculate the exact distance of the toe of the embankment from the centre line at this cross-section.

Solution (see fig. 74)

For method see page 57.

Grade staff reading = 115 − (112·32 + 1·88) = 1·80.

If toe of bank intersects ground at A then $(12 - 5) \times \frac{2}{3} = 4{\cdot}89 + 0{\cdot}80$
$4{\cdot}67 = 5{\cdot}69.$

If toe of bank intersects ground at B then $(15 - 5) \times \frac{2}{3} = 5{\cdot}18 + 0{\cdot}80$
$6{\cdot}67 = 5{\cdot}98.$

From the above calculations it is seen that at A the calculated level is less than the actual level, and at B it is greater; therefore the toe intersects the ground between A and B.

Consider triangles A_1A_2C and B_1B_2C (these are similar triangles).

To find the horizontal distance AC.

The distance AB = 15 − 12 = 3 m

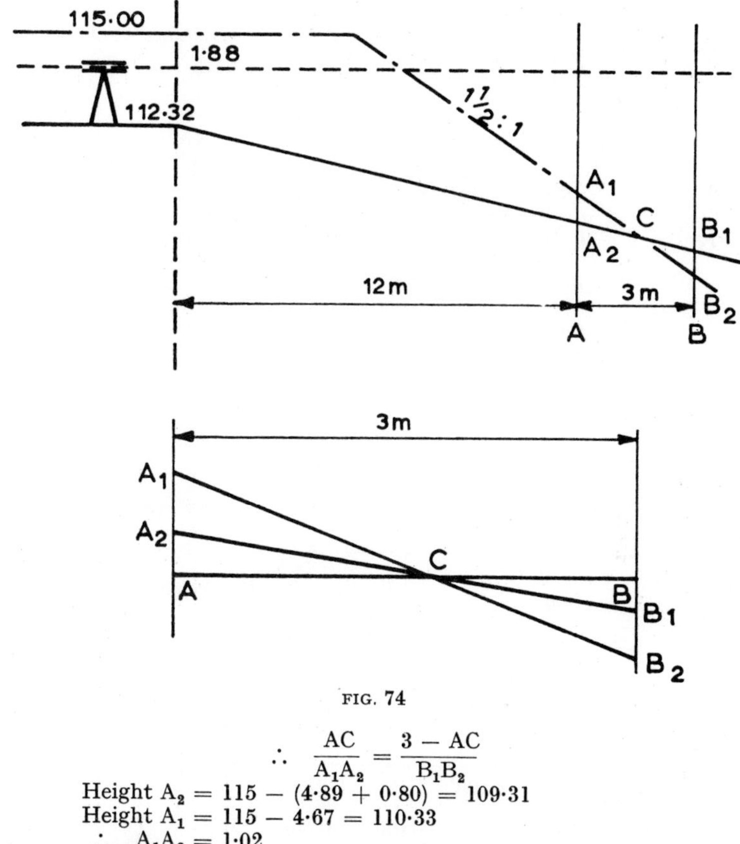

FIG. 74

$$\therefore \quad \frac{AC}{A_1A_2} = \frac{3 - AC}{B_1B_2}$$

Height $A_2 = 115 - (4{\cdot}89 + 0{\cdot}80) = 109{\cdot}31$
Height $A_1 = 115 - 4{\cdot}67 = 110{\cdot}33$
$\therefore \quad A_1A_2 = 1{\cdot}02$
Height $B_1 = 115 - (5{\cdot}18 + 0{\cdot}80) = 109{\cdot}02$
Height $B_2 = 115 - 6{\cdot}67 = 108{\cdot}33$
$\therefore \quad B_1B_2 = 0{\cdot}69$

$$\therefore \quad AC = \frac{3 \times A_1A_2}{A_1A_2 + B_1B_2} = \frac{3 \times 1{\cdot}02}{1{\cdot}02 + 0{\cdot}69} = \frac{3{\cdot}06}{1{\cdot}71} = 1{\cdot}79$$

The exact distance from the centre line to the toe of the embankment $= 12{\cdot}0 + 1{\cdot}79 = 13{\cdot}79$ metres.

Example

Prepare a level book to show the rises and falls required to set up sight rails for a sewer at three manhole positions X, Y and Z. The readings on the staff at a near-by bench mark, value 67·841 was 0·310; staff readings at ground level were X, 1·327; Y, 1·214; Z, 1·607. The invert level of the sewer at X is to be 65·248.

Gradient X to Y, 1 in 80 distance 40 m.

Gradient Y to Z, 1 in 100 distance 60 m.

The boning rods are 3 m long. What is the depth of each manhole?

Back-sight	*Inter. sight*	*Fore-sight*	*Rise*	*Fall*	*Reduced Level*	*Distance*	*Remarks*
0·310					67·841		OBM
	1·327			1·017	66·824	0	G.L. at X.
	1·214		0·113		66·937	40	G.L. at Y.
	1·607			0·393	66·544	100	G.L. at Z.
		0·310	1·297		67·841	0	OBM
0·310		0·310	1·410	1·410	67·841		
		0·310		1·410	0·000		
		0·000		0·000			

$$\text{Fall from X to Y} = \frac{40 \times 1}{80} = 0{\cdot}50 \text{ m}$$

$$\text{Invert level at Y} = 65{\cdot}248 - 0{\cdot}50 = 64{\cdot}748 \text{ metres}$$

$$\text{Fall from Y to Z} = \frac{60 \times 1}{100} = 0{\cdot}60 \text{ m}$$

$$\text{Invert level at Z} = 64{\cdot}748 - 0{\cdot}60 = 64{\cdot}148 \text{ metres.}$$

Back-sight	*Inter-sight*	*Fore-sight*	*Rise*	*Fall*	*Reduced Level*	*Distance*	*Remarks*
					67·841		OBM
			0·407		68·248	0	Sight Rail at X.
				0·500	67·748	40	Sight Rail at Y.
				0·600	67·148	100	Sight Rail at Z.
			0·693		67·841		OBM
			1·100	1·100	67·841		
				1·100	0·000		
				0·000			

Depth of manhole at X = 66·824 − 65·248 = 1·576.
Depth of manhole at Y = 66·937 − 64·748 = 2·189.
Depth of manhole at Z = 66·544 − 64·148 = 2·396.

Example

In a highway revision project a new road about 1½ miles long is required to leave the existing road, by-pass a village, and rejoin the old road, and the only map available is an up-to-date 6″ Ordnance Survey Map showing contours at 50′ vertical interval. The country surrounding the village is gently undulating containing only a few scattered farm buildings.

Indicate briefly the survey procedure to be adopted to provide the plans and sections necessary to carry out the project. List these plans together with suggested suitable scales and state what surveying instruments would be required.

(Inst. Civ. Engs. Sec. B, October 1949.)

Example

On a railway the formation level at peg 120 + 00 was 152·20 m, and the gradient 1 in a 100 up, followed by a down gradient of 1 in 200, the reduced level at peg 138 + 00 on the down gradient being 155·20 m. The two gradients, both in cutting, originally had no connecting vertical curve, but it was decided to introduce a parabolic one 1000 m in length. Calculate the extra depth of cutting required for this purpose at the intersection point of the gradients and at peg 126 + 00.

Answer: 1·875 m and 0·675 m.

4 AREAS AND VOLUMES

THE following formulæ are of considerable importance in calculating the solutions of various problems connected with surveying and levelling.

The Triangle

1. Given base and perpendicular height (see fig. 75).

$$\text{Area} = \frac{\text{Base} \times \text{height}}{2} = \frac{b \times h}{2}$$

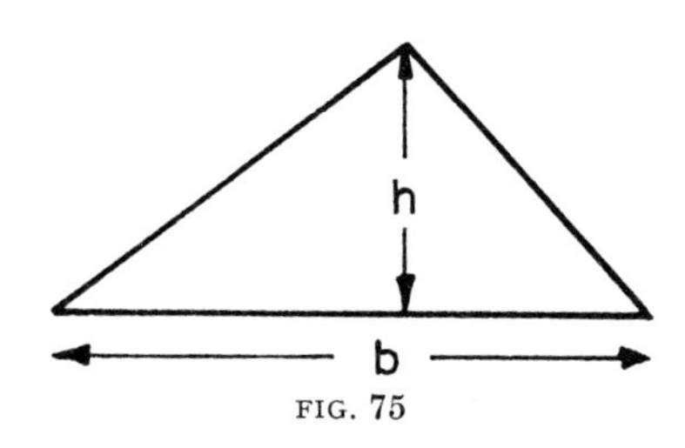

FIG. 75

2. Given the lengths of three sides (see fig. 76).

$$\text{Area} = \sqrt{s(s - a)(s - b)(s - c)}. \quad \text{Where } s = \frac{a + b + c}{2}.$$

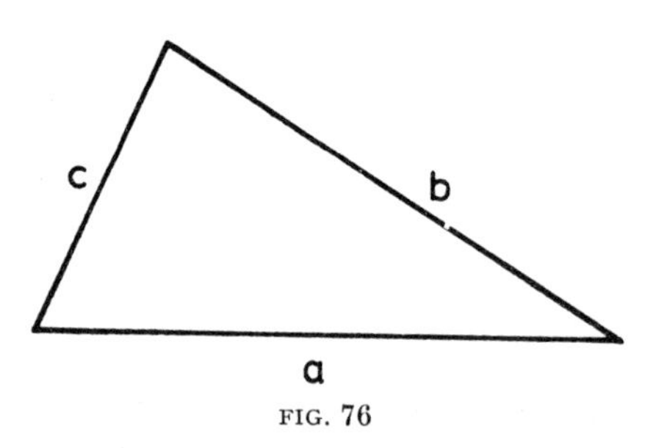

FIG. 76

3. Given two sides and the **included** angle (see fig. 77).

$$\text{Area} = \frac{ab \sin C}{2} = \frac{ac \sin B}{2} = \frac{bc \sin A}{2}.$$

Note—Formula 2 is generally used, as in surveying problems the lengths of the sides are usually known.

Example

A triangular plot has sides with the following lengths: 1000 m, 1300 m and 700 m. Calculate the area of the plot in hectares.

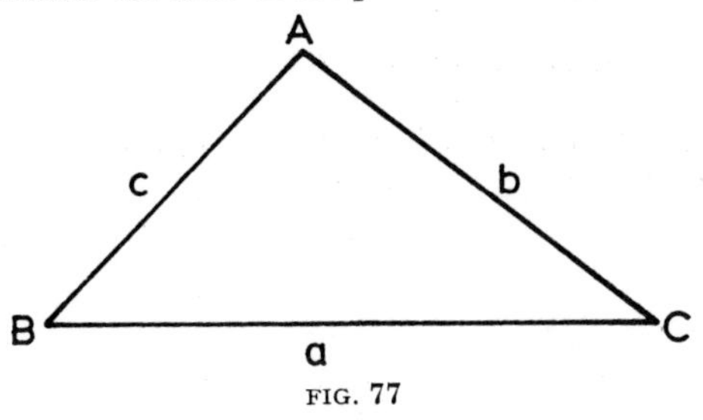

FIG. 77

Solution (use formula 2)

$$s = \frac{a + b + c}{2} = \frac{1000 + 1300 + 700}{2} = 1500.$$

$(s - a) = 500.$ $(s - b) = 200.$ $(s - c) = 800.$

$$\begin{aligned} \text{Area} &= \sqrt{s(s - a)(s - b)(s - c)} \\ &= \sqrt{1500 \times 500 \times 200 \times 800} \text{ m}^2 \\ &= \frac{\sqrt{1500 \times 500 \times 200 \times 800}}{100 \times 100} \text{ hectares} \\ &= 34{\cdot}62 \text{ hectares.} \end{aligned}$$

Note.—The formula used in the worked example is suitable for logarithmic calculation and therefore easily manipulated.

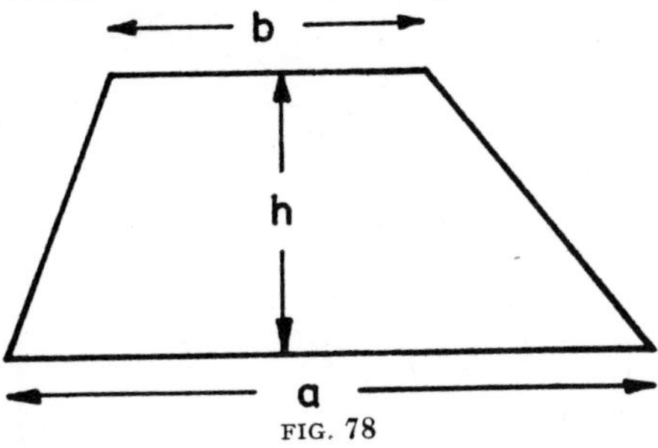

FIG. 78

QUADRILATERALS

1. **The Trapezium** (see fig. 78).

$$\begin{aligned} \text{Area} &= \text{Half sum of parallel sides} \times \text{height} \quad \text{or} \\ &= \text{Half height} \times \text{sum of parallel sides} \\ &= \frac{(a + b) \times h}{2}. \end{aligned}$$

2. **The Parallelogram** (see fig. 79).

$$\begin{aligned} \text{Area} &= \text{Base} \times \text{perpendicular height} \\ &= a \times h. \end{aligned}$$

If the height is not known but the sides and any angle is known, the formula for the area is:

Area $= ab \sin A$ or
$= ab \sin B$.

Note that $\sin A = \sin B$ because the two angles are supplementary.

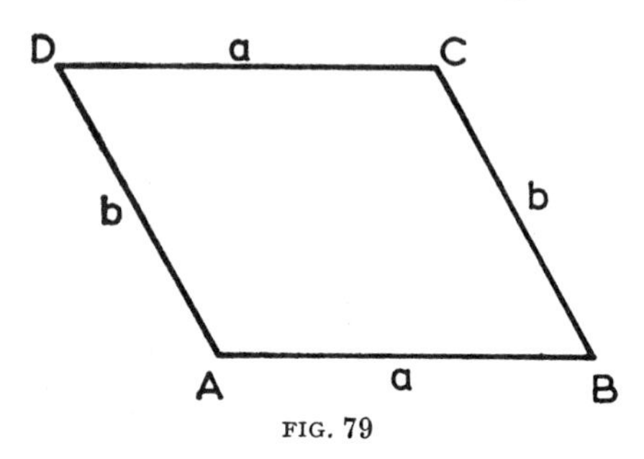

FIG. 79

3. The Polygon (see fig. 80).

Split up the figure into triangles by joining up corner points.
Area = ABC + ACD + ADE.

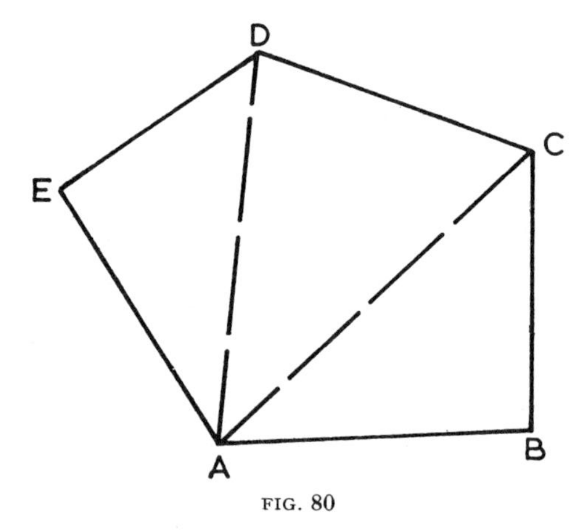

FIG. 80

IRREGULAR FIGURES WITH CURVED BOUNDARIES

1. Simpson's Rule (see fig. 81).

The base line must be divided into an **even** number of equal intervals. This will give an **odd** number of heights or ordinates as shown in figure.

$$\text{Area} = \frac{\text{Interval}}{3}(\text{Sum of end ordinates} + \text{twice sum of odd ordinates} + \text{four times sum of even ordinates}).$$

$$\text{Area of ABCD} = \frac{x}{3}\left[(y_1 + y_7) + 2(y_3 + y_5) + 4(y_2 + y_4 + y_6)\right].$$

To use Simpson's Rule, an **odd** number of ordinates is essential. If an even number of ordinates is given, neglect the first one, calculate by Simpson's Rule and then add the neglected area calculated by the trapezoidal rule (see below).

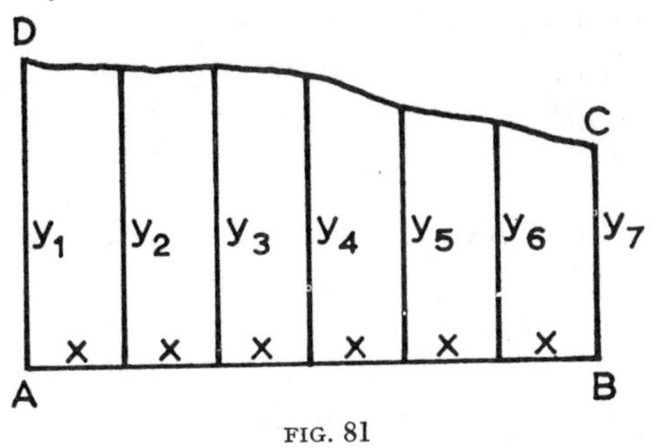

FIG. 81

2. **Trapezoidal Rule** (see fig. 82).

Used when there are **any** number of heights or ordinates.

Area = Interval (Half sum of first and last ordinates + remaining ordinates).

$$\text{Area of ABCD} = x\left[\left(\frac{y_1 + y_8}{2}\right) + y_2 + y_3 + y_4 + y_5 + y_6 + y_7\right].$$

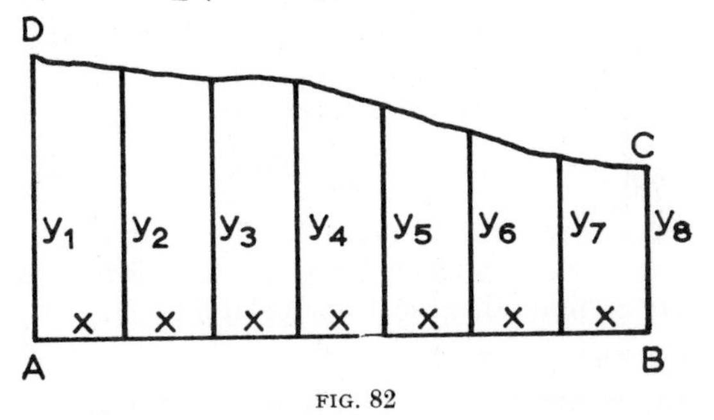

FIG. 82

Example

A strip of land is 960 metres long. This length is marked off into eight equal intervals, and the consecutive breadths were measured at the ends of the intervals as follows:

5, 13, 15, 18, 20, 24, 12, 6 and 5 metres.

Calculate the area in hectares.

Solution

$$\text{Each interval} = \frac{960}{8} = 120 \text{ metres}$$

Sum of first and last ordinates = 10.
Twice sum of all other odd ordinates = 2(15 + 20 + 12) = 94.
Four times sum of all other even ordinates = 4(13 + 18 + 24 + 6) = 244.

$$\text{Area} = \frac{120}{3}\left(10 + 94 + 244\right) \quad \frac{120}{3} \times 348 \text{ m}^2$$
$$= 13\,920 \text{ m}^2$$
$$= 1{\cdot}392 \text{ hectares.}$$

3. Trapezoidal Formula for Volumes

A trapezoid is a solid figure with two parallel plane end areas which are trapezia. A typical example of a trapezoid is the volume of a cutting between two cross-sections where the floor of the cutting has a different gradient to the ground level.

Volume = Interval (Half the sum of the end areas + remaining areas)
The volume of such a section is given by:

$$x\left(\frac{y_1}{2} + y_2 + y_3 \ldots + y_6 + \frac{y_7}{2}\right)$$

where y_1 and y_2 etc. are the areas of the sections.
x is the axial distance between two sections.

4. Simpson's Rule for Volumes (Prismoidal Formula)

Cross-section areas take the place of ordinates in the formula for calculating areas with curved boundaries.

It is necessary to have an **odd** number of cross-section areas at equal intervals.

Volume = $\frac{\text{Interval}}{3}$ (Sum of end cross-section areas + twice sum of odd cross-sections + four times sum of even cross-sections).

The volume of such a section is given by:

$$\frac{x}{3}\left[y_1 + y_7 + 2(y_3 + y_5) + 4(y_2 + y_4 + y_6)\right].$$

Note: Areas and volumes calculated by the two formulae given above will not necessarily be the same. Simpson's Rule is normally accepted as being more accurate than the trapezoidal formula.

CUTTINGS AND EMBANKMENTS

1. Cutting or Embankment on Sloping Ground (see fig. 83).

To find the area of cross-section ABCD and half widths; given side slope, ground slope, formation width, height at centre.

Let: Formation width AB = $2a$
Height at centre = h
Side slope = s:1 (1 vertical)
Ground slope = g:1 (1 vertical)

To find half widths CG and DH (W_1 and W_2):
Erect perpendiculars CX and DY on line AB produced.

$$\text{Then CX} = \text{GO} = \frac{\text{BX}}{s} = \frac{W_1 - a}{s}$$

$$\therefore \quad \text{GM} = \text{GO} - h = \frac{W_1 - a}{s} - h.$$

$$\text{But GM} = \frac{W_1}{g}$$

$$\therefore \quad \frac{W_1}{g} = \frac{W_1 - a}{s} - h = \frac{W_1 - a - hs}{s}$$

$$\therefore \quad W_1 s = W_1 g - ag - hsg$$

$$\therefore \quad W_1(g - s) = ag + hsg$$

$$\therefore \quad W_1 = \frac{g(a + hs)}{(g - s)}.$$

Similarly it can be shown that

$$W_2 = \frac{g(a + hs)}{(g + s)}.$$

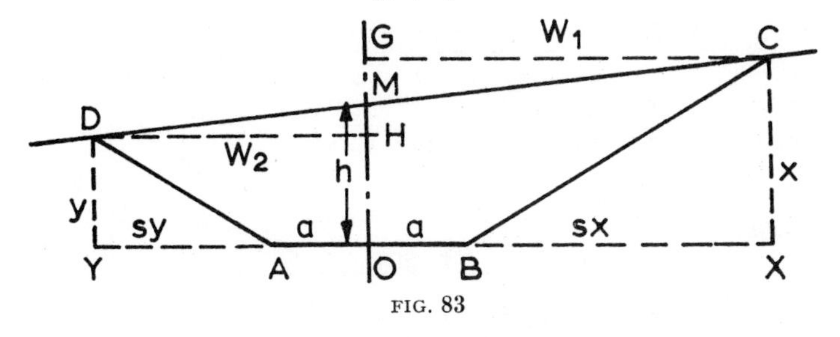

FIG. 83

To find area ABCD:

$$\text{Section ABCD} = \text{CXYD} - \text{CXB} - \text{DYA}$$

$$= \text{XY}\left[\frac{\text{CX} + \text{DY}}{2}\right] - \tfrac{1}{2}\text{CX.BX} - \tfrac{1}{2}\text{DY.AY}.$$

$$\text{Now CX} = \frac{\text{BX}}{s} = \frac{W_1 - a}{s}$$

$$\text{and DY} = \frac{\text{AY}}{s} = \frac{W_2 - a}{s}$$

$\therefore$ Section ABCD

$$= \left(\frac{W_1 + W_2}{2}\right)\left(\frac{W_1 - a}{s} + \frac{W_2 - a}{s}\right) - \tfrac{1}{2}\left(\frac{W_1 - a}{s}\right)(W_1 - a) - \tfrac{1}{2}\left(\frac{W_2 - a}{s}\right)(W_2 - a)$$

$$= \left(\frac{W_1 + W_2}{2}\right)\left(\frac{W_1 - a + W_2 - a}{s}\right) - \frac{1}{2}\left(\frac{(W_1 - a)^2}{s} + \frac{(W_2 - a)^2}{s}\right)$$

$$= \frac{1}{2s}\left((W_1 + W_2)(W_1 + W_2 - 2a) - (W_1 - a)^2 - (W_2 - a)^2\right)$$

$$= \frac{1}{2s}\left(W_1^2 + W_1W_2 - 2aW_1 + W_2W_1 + W_2^2 - 2aW_2 - W_1^2 \right.$$
$$\left. - a^2 + 2aW_1 - W_2^2 - a^2 + 2W_2a\right)$$

$$= \frac{1}{2s}\left(2W_1W_2 - 2a^2\right)$$

$$= \frac{W_1W_2 - a^2}{s}$$

$$\therefore \quad \text{Area of section ABCD} = \frac{W_1W_2 - a^2}{s}.$$

Note a in the formula is **half** the formation width.

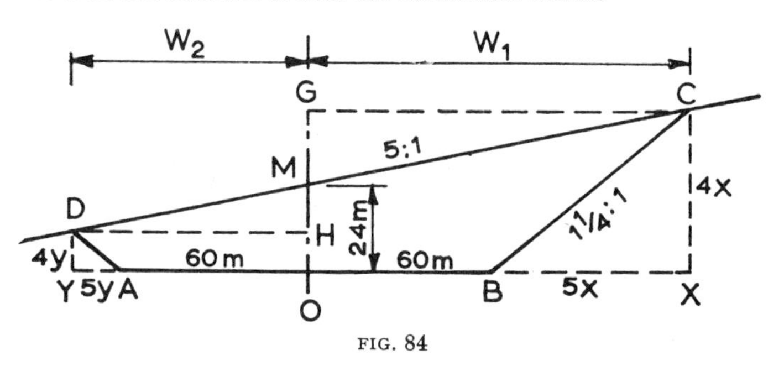

FIG. 84

Example (see fig. 84)

Given: Formation width = 120 m
Height at centre = 24 m
Side slopes = $1\frac{1}{4}:1$
Ground slope = 5:1

$$W_1 = \frac{g(a + hs)}{g - s} = \frac{5(60 + 30)}{3\frac{3}{4}} = 5 \times 90 \times \frac{4}{15} = 120 \text{ m}$$

$$W_2 = \frac{g(a + hs)}{g + s} = \frac{5(60 + 30)}{6\frac{1}{4}} = 5 \times 90 \times \frac{4}{25} = 72 \text{ m}$$

$$\text{Area} = \frac{W_1W_2 - a^2}{s} = \frac{120 \times 72 - 60^2}{1\frac{1}{4}} = \frac{8640 - 3600}{1\frac{1}{4}} = 5040 \times \frac{4}{5}$$
$$= 4032 \text{ m}^2.$$

Note.—If instead of the side slope and ground slope being given as such. the angles which sides and ground make with the horizontal are given. then the cotangents of these angles will give the correct values of s and g in the above formulae.

Above Example Worked from First Principles

$$GC = 5GM \quad GC = 60 + 5x \quad GM = 4x - 24$$

$$\therefore \quad 60 + 5x = 5(4x - 24)$$

$$\therefore \quad 15x = 180$$

$$\therefore \quad x = 12 \quad CX = 48 \quad BX = 60.$$

Also $DH = 5HM$.

Now $DH = 60 + 5y$

and $HM = 24 - 4y$

$$\therefore \quad 60 + 5y = 5(24 - 4y) = 120 - 20y$$

$$\therefore \quad 25y = 60 \qquad \therefore \quad y = 2{\cdot}4$$

and $DY = 9{\cdot}6 \qquad AY = 12.$

$$\text{Area } ABCD = DYXC - DYA - BXC$$

$$= \left(\frac{48 + 9{\cdot}6}{2}\right)(120 + 60 + 12) - \frac{9{\cdot}6 \times 12}{2} - \frac{48 \times 60}{2}$$

$$= \left(\frac{57{\cdot}6}{2} \times 192\right) - 57{\cdot}6 - 1440$$

$$= 5529{\cdot}6 - 57{\cdot}6 - 1440$$

$$= 4032 \text{ m}^2.$$

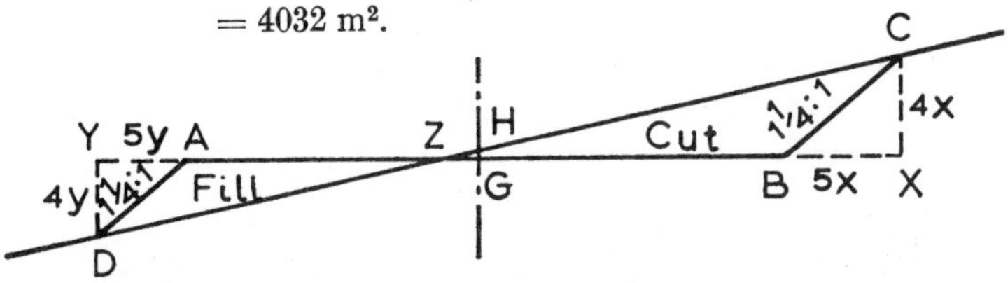

FIG. 85

2. **Cut and Fill Section** (see fig. 85).

It is considered that it is easier to calculate this cross-sectional area from first principles rather than from formula.

Given: Formation width $AB = 40$ m

Height at centre $= 0{\cdot}5$ m

Side slopes $= 1\frac{1}{4}{:}1$

Ground slope $= 10{:}1$.

Draw perpendiculars CX and YD from C and D on to AB produced.

Let $BX = 5x$, then $CX = 4x$ (slope $1\frac{1}{4}{:}1$).

Similarly, $DY = 4y$ and $AY = 5y$.

$$\text{Area of cut} = \text{Area of triangle } ZBC = \frac{ZB \times CX}{2}.$$

Since $HG = 0{\cdot}5$ m then $ZG = 5$ m (10:1 ground slope)

$$\therefore \quad ZB = 25 \text{ m}.$$

Also $ZX = 10CX$ (10:1 ground slope)

$$\therefore \quad ZB + BX = 10CX$$

i.e. $25 + 5x = 40x$.

$$x = \frac{25}{35} = \frac{5}{7}.$$

$$CX = \frac{20}{7}.$$

$$BX = \frac{25}{7}.$$

$$\text{Area of cut} = \frac{ZB \times CX}{2} = \frac{25}{2} \times \frac{20}{7} = \frac{250}{7} = 35\tfrac{5}{7}\ \text{m}^2.$$

$$\text{By similar triangles } \frac{\text{Area ADZ}}{\text{Area ZBC}} = \frac{AZ^2}{ZB^2}$$

$$\therefore \quad \text{Area of fill (triangle ADZ)} = \frac{15^2}{25^2} \times \frac{35\tfrac{5}{7}}{1} = \frac{225}{625} \times \frac{250}{7}$$

$$\text{Area of fill} = 12{\cdot}9\ \text{m}^2.$$

THE MASS-HAUL CURVE

The cost of excavating soil depends not only upon the volume of material excavated but also upon its disposal.

In the case of a road or railway, it is common practice to use the material excavated from cuttings for forming embankments.

For economic construction the 'cut' should balance the 'fill' (i.e. the volume of earth excavated from cuttings should equal the volume of earth required for the embankments), and the haul length from cut to fill should be as short as possible. The mass-haul curve shows graphically the haul length, the economic direction of haul, and the position along the route where the cut balances the fill.

Method of Constructing the Mass-haul Curve

1. Calculate cross-sectional areas at convenient intervals.
2. Calculate the volume between adjoining cross-sections, letting cut volumes be positive and fill volumes be negative.
3. Sum the volumes algebraically from length to length.
4. Draw the base line under the longitudinal section, and mark off chainage to the same scale as the longitudinal section.
5. Plot the algebraic sum of the volume vertically against the length horizontally.

Example

Plot the mass-haul curve for a 10 m wide road with side slopes 1 in 1, whose longitudinal section is as shown in fig. 86, the road being level.

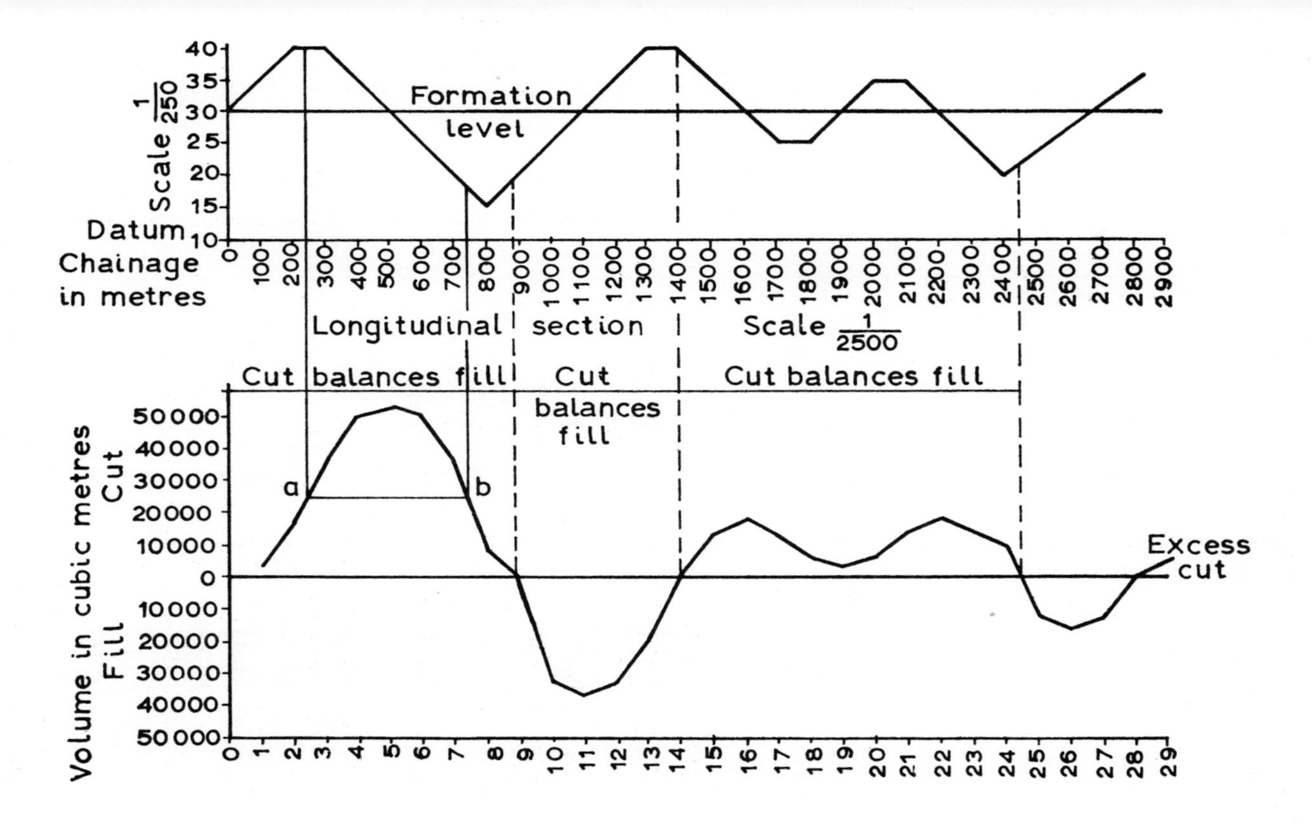

Formation level
Scale 1/250
Datum
Chainage in metres
Longitudinal section
Scale 1/2500
Cut balances fill
Cut balances fill
Cut balances fill
a
b
Volume in cubic metres
Cut
Fill
Excess cut

FIG. 86

CALCULATION FOR MASS-HAUL CURVE

Section	*Cross-sectional Area*	*Volume*	*Algebraic Sum of Volume*
0	0		0
		+3333	
1	+75		+3333
		+13 333	
2	+200		+16 666
		+20 000	
3	+200		+36 666
		+13 333	
4	+75		+49 999
		+3333	
5	0		+53 333
		−3333	
6	−75		+49 999
		−13 333	
7	−200		+36 666
		−28 333	
8	−375		+8333
		−28 333	
9	−200		−20 000
		−13 333	
10	−75		−33 333
		−3333	
11	0		−36 666
		+3333	
12	+75		−33 333
		+13 333	
13	+200		−20 000
		+20 000	
14	+200		0
		+13 333	
15	+75		+13 333
		+3333	
16	0		+16 666
		−3333	
17	−75		+13 333
		−7500	
18	−75		+6833
		−3333	
19	0		+3500
		+3333	
20	+75		+6833
		+7500	
21	+75		+14 333
		+3333	
22	0		+17 666
		−3333	
23	−75		+14 333

CALCULATION TABLE FOR MASS-HAUL CURVE—*continued*

Section	*Cross-sectional Area*	*Volume*	*Algebraic Sum of Volume*
		−13 333	
24	−200		+1000
		−13 333	
25	−75		−12 333
		−3333	
26	0		−15 666
		+3333	
27	+75		−12 333
		+13 333	
28	+200		+1000
		+4333	
29	0		+5333

Notes

It will be noted that:

1. When the curve slopes upwards, the work is cutting.
2. When the curve slopes downwards, the work is excavation.
3. The maximum points on the curve coincide with the ends of the cuts.
4. The minimum points on the curve coincide with the ends of the fills.
5. The curve starts at zero, and when it again reaches zero, the cut balances the fill between the two points.
6. Any horizontal line, such as *ab*, drawn across the curve will give cut and fill balance between the two points of intersection *a* and *b*.
7. If at the end of the route the curve is below zero, then fill exceeds cut and extra material will have to be brought on to the site.
8. If, however, the curve is above zero, then cut exceeds fill and the excess of excavated material will have to be carted from the site.
9. To avoid either of these difficulties, the proposed level of the route may be adjusted so that cut and fill balance.

If this cannot be done then, where there is insufficient fill it is usually cheaper to make the cut wider than to bring in extra material. Where there is excess fill, it is cheaper to widen the fill than to transport earth to a dumping ground off the site.

Economic haul length is reached when it is cheaper to dump excavated material on to a spoil heap and bring in material from a borrow pit for fill.

10. No allowance has been made for 'shrinkage' of the excavated material, which may vary from 0 to 10%, depending on the soil and construction methods. When the bulking percentage is known, the positive volumes are decreased by the percentage.

Example

A breakwater, trapezoidal in cross-section, is 150 m long. At the shore end, the section is as follows: depth 12 m, top width 10 m, each sloping side being battered or sloped 1 horizontal to 12 vertical (i.e. bottom width is 12 m). At the outer end the depth is 18 m and the top width 10 m. At all sections of the wall the batter is constant as above, the top face is horizontal, and the depth of the wall changes uniformly. Find the volume of material in the wall.

Solution

$$\text{Cross-sectional area of shore end} = \frac{10 + 12}{2} \times 12$$
$$= 132 \text{ m}^2$$
$$\text{Cross-sectional area of outer end} = \frac{10 + 13}{2} \times 18$$
$$= 207 \text{ m}^2$$
$$\text{Cross-sectional area of centre section} = \frac{10 + 12 \cdot 5}{2} \times 15$$
$$= 168 \cdot 75 \text{ m}^2$$
$$\text{Volume} = x\left(\frac{y_1}{2} + y_2 + \frac{y_3}{2}\right)$$
$$= 75\left(\frac{132}{2} + 168 \cdot 75 + \frac{207}{2}\right)$$
$$= 25\,369 \text{ m}^3.$$

Example

A railway cutting has a formation width of 10 m and the side slopes are 1 vertical to 1·5 horizontal. The ground surface is everywhere horizontal. The depths of the cutting at the centre line are given in the table:

Distance (metres) .	0	50	100	150	200	250	300
Depth (metres) .	6·64	6·64	7·2	8·6	8·4	9·2	10·4

Find the volume of excavation in cubic yards over this length of cutting, using the prismoidal formula.

Example

A tank, circular in plan, has an inside diameter of 100 m. The wall, trapezoidal in cross-section, is 11 m high and has a vertical water face. If the top and bottom widths are 1·5 m and 5·33 m respectively, calculate the volume of concrete required for the walls and also the weight, taking the density as 2307 kg/m³.

Solution

When the centre line is curved, the volume of the solid is equal to the area multiplied by the length of the path traced by the centre of gravity of the area.

To find the distance of centre of gravity from inner face, take moments of area about the inner face.

$$1{\cdot}5\times 11\times\frac{1{\cdot}5}{2}+(5{\cdot}33-1{\cdot}5)\times\frac{11}{2}\times\left(\frac{3{\cdot}83}{3}+1{\cdot}5\right)$$

$$=1{\cdot}5\times 11+\frac{5{\cdot}33-1{\cdot}5}{2}\times 11\times\overline{X}$$

$$1{\cdot}5\times 11\times 0{\cdot}75+3{\cdot}83\times 5{\cdot}5\times 2{\cdot}777=3{\cdot}415\times 11\times\overline{X}$$

$$12{\cdot}375+58{\cdot}497=37{\cdot}565\overline{X}$$

$$\overline{X}=\frac{70{\cdot}872}{37{\cdot}565}=1{\cdot}88.$$

Diameter across centre of gravity $= 100 + (2 \times 1{\cdot}88)$
$= 103{\cdot}76.$

Volume $= 2\pi \times 51{\cdot}88 \times 37{\cdot}565$
$= 12\,244{\cdot}8 \text{ m}^3$

Weight $= 2307 \times 12\,244{\cdot}8$
$= 28\,248\,754$ kg.

Example

The straight centre line of a proposed railway is located by 100 m chainage pegs, and cross-sections have been taken at these pegs.

From the longitudinal section, the formation level is in cut from chainage 76 + 60 m to chainage 83 + 78 m. From the cross-sections, the areas of excavation down to formation level at the intervening pegs are:

Chainage of Peg (m)	*Area of Excavation (m^2)*
77 + 00	16
78 + 00	304
79 + 00	656
80 + 00	568
81 + 00	640
82 + 00	320
83 + 00	176

Estimate the volume of excavation in m^3 involved in the cutting, using (*a*) the trapezoidal formula and (*b*) the prismoidal formula.

Answer: (*a*) 265 584 m^3; (*b*) 258 917 m^3.

Example

The following are particulars of a length of proposed road. Measurements in metres.

Chainage	*Formation Level*	*Ground Level*
400	70·12	79·14
500		78·23
600		77·61

The formation width is 30 m; side slopes are 1½ horizontal to 1 vertical, gradient 1 in 30 falling from chainage 400. Cross fall of ground looking in direction of chainage: 6½°, 7°, 7½°, at 400 m, 500 m and 600 m respectively.

Determine the volume of cutting for the length in cubic metres.

5 THE MEASUREMENT OF ANGLES

INSTRUMENTS

THE BOX SEXTANT

THIS instrument is similar in construction to the optical square, differing from it in that one of the mirrors is pivoted, and has a scale and pointer attached to it so that angles up to about 120° may be measured. A telescope is generally attached to the eyepiece to enable long sights to be taken.

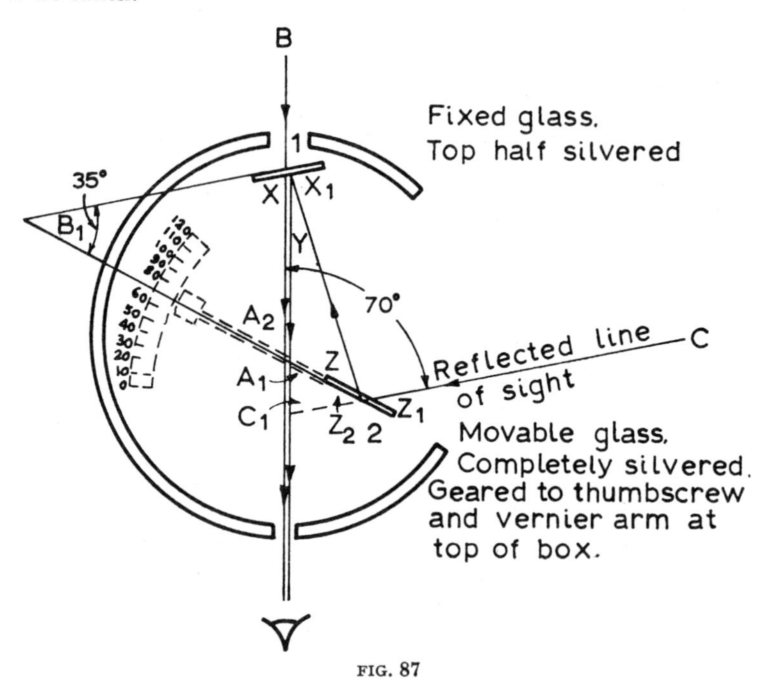

FIG. 87

Its design is based upon the fact that the change of direction of a ray of light reflected from two mirrors is twice that of the angle between them (see fig. 87). The observer at station A views a pole at station B through the bottom half of mirror No. 1, whilst at the same time the image of the pole at station C is reflected into the top half from mirror No. 2. When the two images have been made to coincide by turning the pivoted mirror No. 2, the angle BC_1C is recorded on the scale.

Thus, given that a ray of light leaves the surface of a mirror at the same angle as that at which it strikes the mirror, then angle X = angle X_1 and angle Z = angle Z_1. Also angles A_2 and A_1 are equal and Z_1 and Z_2 are equal, being opposite angles.

Angles $X + Y + X_1 = 180°$ and angles $Y + Z + Z_2 + C_1 = 180°$.

Therefore $X + Y + X_1 = Y + Z + Z_2 + C_1$

or $2X + Y = 2Z + Y + C_1$

And as Y is common to both sides, then

$$2X - 2Z = C_1 \text{ or } X - Z = \frac{C_1}{2}.$$

Also triangle B_1XA_2 and triangle $C_1A_1Z_2$ have equal angles and, as A_2 is common to both sides, $B_1 + X - Z_1 = C_1$. But $X - Z_1 = \frac{C_1}{2}$.

Therefore, subtracting $\frac{C_1}{2}$, $B_1 = C_1 - \frac{C_1}{2} = \frac{C_1}{2}$.

Thus the angle B_1 between the mirrors is always half the angle of sight C_1.

The sextant measures the actual angle between stations so that if the stations are at considerably differing levels the angle recorded must be reduced mathematically to the horizontal angle required.

PRISMATIC COMPASS

This instrument consists of:

1. A magnetised needle mounted in a circular metal box so that it may rotate freely, and fixed to it a circular card graduated into 360°.

The needle is either balanced on a fine pivot, in which case there is a device to lift it off when not in use, or the needle is floating in a liquid.

2. Two hinged sighting arms hinged to the box, one of which has a prism mirror attached so that the observer may view the card without removing his eye from the instrument. The arm opposite to the eye-piece is provided with a hair line for sighting purposes. It should be noted that the graduations are commenced from the **south** end of the card so that the observer may read directly the angle measured (see fig. 88).

METHOD OF USE

As the compass needle always points north, all angles are measured from the north–south line and in a clockwise direction (see fig. 89). From station A the angle made by the line AB is 40° east of north.

To prevent the needle card from swinging, a brake, operated by a press stud, is provided, which enables the swinging of the card to be slowed down quickly. The instrument allows angles to be measured to

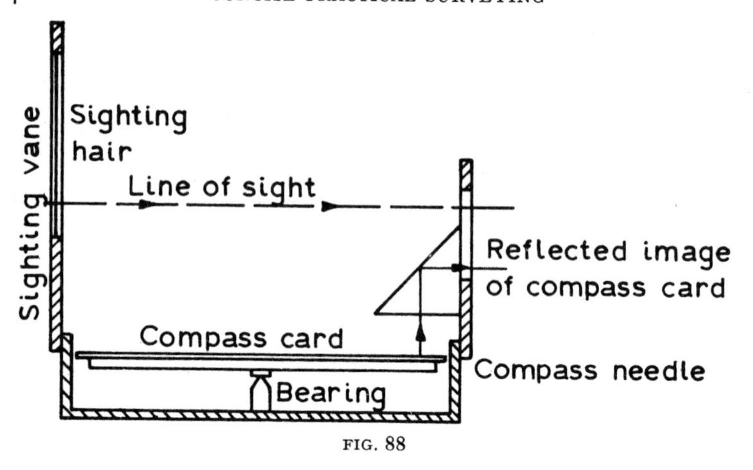

FIG. 88

an accuracy of about $\frac{1}{2}°$. Some compasses are provided with a tripod which allows readings to be taken to a greater degree of accuracy.

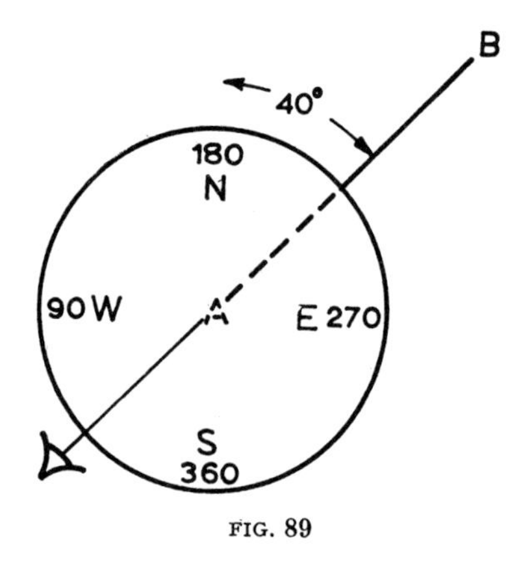

FIG. 89

VERNIER

This device may be used upon any instrument recording measurements on a graduated scale, and in particular on theodolites (see page 85) and tacheometers. It enables the scale to be read to a finer degree of accuracy than would be possible by eye alone. The attachment functions as follows: The vernier scale is divided into 10 divisions which equal 9 divisions on the main scale of the instrument. Therefore 1 division on the vernier $= \frac{9}{10}$ divisions on the main scale. If the arrow or 0 on the vernier is set against a division on the main scale and then moved forward $\frac{1}{10}$ of a division of the main scale, the first graduation on the vernier scale will line in with the next or first division on the main scale. If the vernier arrow is moved $\frac{6}{10}$ of a division along the main scale, then the sixth graduation of the vernier will coincide with a graduation on the main scale. Thus the actual figure recorded by the vernier is the last whole division of the main scale plus the value recorded by the vernier, which is the value on the vernier scale at the point where the vernier division and the main scale division coincide. Vernier scales may be divided in several ways; that shown in fig. 90 is divided to read to the nearest minute, the main scale being divided to read to the nearest $\frac{1}{2}$°. The vernier is therefore divided into 30 parts. The value shown recorded in fig. 90 is 65° 9′, as the arrow is placed between 65° and 65° 30′ and the coincidence of lines takes place at 9 on the vernier scale.

THE ABNEY LEVEL

This instrument is generally used to obtain roughly the angle of slope of the ground. It consists (see fig. 91) of a pin-hole telescope with a graduated scale fixed rigidly to it, a **movable** pointer with a level tube attached being fixed to the scale. The bubble tube is not cased, and it may be seen by the observer when looking through the telescope by means of a mirror fixed at an angle in the telescope and covering half the width of the tube. A hole is cut in the top of the tube to admit the bubble image. A vernier is sometimes attached to the pointer. The angle of slope is obtained by viewing a mark on a staff held by an assistant at some distance away, the mark being at the eye-level of the observer. The bubble tube and pointer are then turned until the bubble floats centrally, the angle of slope being then indicated on the scale.

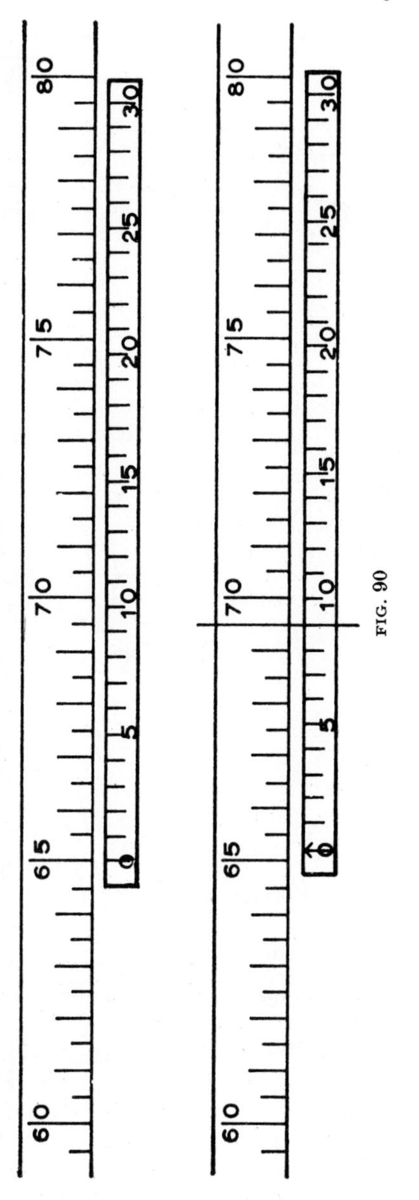

FIG. 90

THE THEODOLITE

This is essentially a telescope whose line of collimation may be revolved through 360° horizontally and transited (i.e. revolved) through

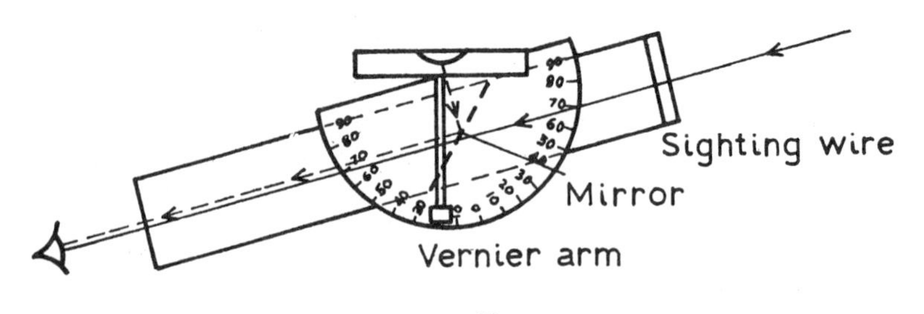
Sighting wire
Mirror
Vernier arm

FIG. 91

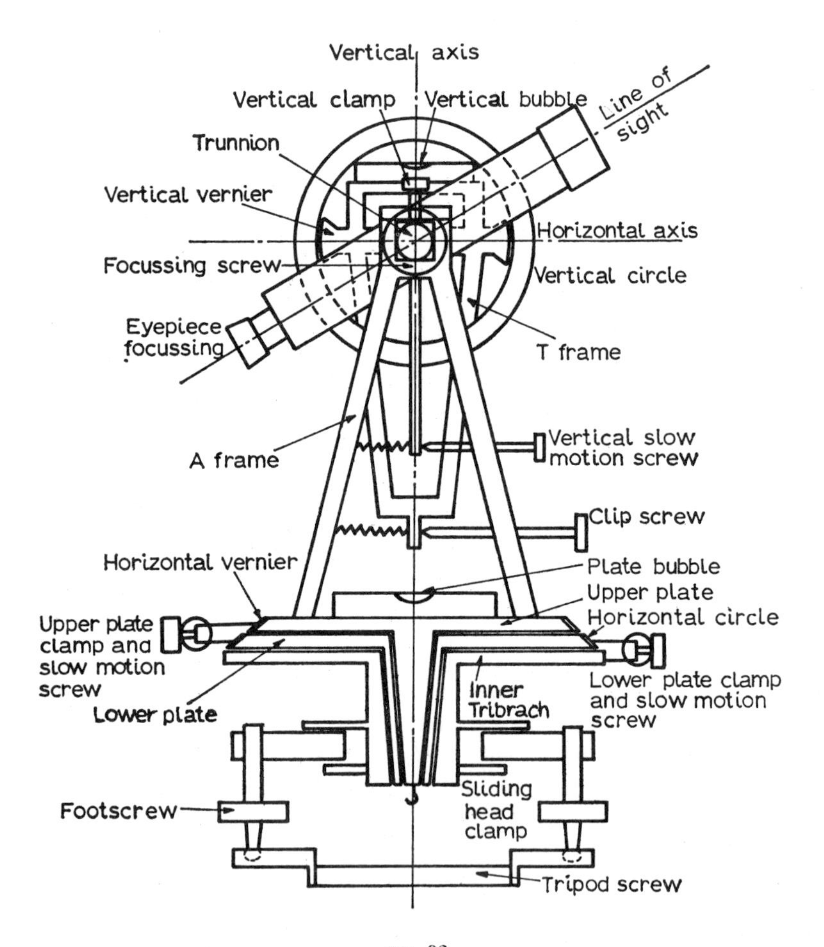
Vertical axis
Vertical clamp
Vertical bubble
Line of sight
Trunnion
Vertical vernier
Horizontal axis
Focussing screw
Vertical circle
Eyepiece focussing
T frame
A frame
Vertical slow motion screw
Clip screw
Horizontal vernier
Plate bubble
Upper plate
Horizontal circle
Upper plate clamp and slow motion screw
Lower plate
Inner Tribrach
Lower plate clamp and slow motion screw
Footscrew
Sliding head clamp
Tripod screw

FIG. 92

360° vertically and the angles of revolution measured (see fig. 92).

It consists of:

1. The Levelling or Tribrach System is similar to that used on a level. A **plate bubble** is attached to the **upper plate** and this is used to set the vertical axis vertical using the **levelling screws.**

2. The Horizontal plates. The **upper plate** carries vernier scales on two opposite edges. These scales read against the **horizontal circle,** graduated in 360° carried on the outside edge of the **lower plate.** The **outer tribrach** is attached to, and can move horizontally in relation to the **Tribrach system** by means of the **sliding head.** A plumb bob is attached by means of a hook set in the vertical axis.

The **lower plate clamp** locks the **lower plate** to the **inner tribrach,** and the **lower plate slow motion screw** allows them to rotate in relation to each other by a small amount under control. The circle is then fixed in position and the **upper plate** with telescope attached rotates in relation to it.

The **upper plate clamp** locks the **upper plate** to the **lower plate,** and the **upper plate slow motion screw** allows them to rotate in relation to each other by a small amount under control. The circle then moves with the **upper plate** and telescope in relation to the **lower plate.**

3. Two **A frames** which are mounted on the **upper plate** supporting the **telescope trunnions** in two bearings.

4. The **telescope** and **vertical circle** are fixed and rotate vertically together. The **vertical circle** may be divided in one of several ways, but in every case 0° or some multiple of 90° defines the horizontal reading and this reading is in a known relationship to the **line of sight or line of collimation.** The vertical rotation is controlled by the **vertical clamp** and the **vertical slow motion screw.** The Telescope is optically the same as the level telescope.

5. The **T frame.** This carries the **vertical bubble** and the two **vertical verniers** and moves under the control of the **clip screw.** The **vertical bubble** is so adjusted that, when it is central, the line joining the zeros of the two **vertical verniers** is in the same relationship to the horizontal as the horizontal reading of the **vertical circle** is to the **line of sight,** so that the vertical angle is read directly from the circle.

Note—It is essential that a theodolite be available when dealing with the construction and use of this instrument.

SETTING UP OR TEMPORARY ADJUSTMENT

The object is to set the vertical axis vertical and passing through the station mark. The plumb bob assists in this since, when it is steady, it hangs in the position the vertical axis will assume when it is vertical.

1. Take the instrument from its box, making sure you can put it

back again. Screw it onto the tripod, lock all the clamps and centralise the screws.

2. Place the complete instrument over the station mark so that the plumb bob point is within 50 mm or less of it, and the tripod head is approximately level.

3. Move the legs singly until the plumb bob point is within 10 mm or less of the mark. The plumb bob moves in the same direction as the leg which is moved. On soft ground the feet of the tripod must be trodden into the ground for greater stability, this will move the plumb bob towards the leg being trodden in.

4. Using the sliding head, set the plumb bob point exactly over the ground mark and lock the clamps.

5. Level the instrument with the footscrews and plate bubble in a similar manner to the level (see page 27).

6. Eliminate parallox using the eyepiece and focusing screws in a similar manner to the level (see page 27).

FACE LEFT AND FACE RIGHT

All angles must be measured twice, once on face left and once on face right. Face left is when the vertical circle is to the left of the telescope, viewed from the eyepiece, face right when it is to the right. To change from face left to face right, rotate the telescope 180° horizontally then 180° vertically. Readings on face left are thus read on the opposite sides of the circles to those on face right.

The reasons for this double reading are:

(*a*) It provides a check on gross errors.

(*b*) Two values for the angle are found, one of which will be too large, the other too small, both by the same amount called the **Collimation Error.** Hence the mean will be correct with the error eliminated. The Collimation Error arises from small maladjustments in the instrument.

MEASURING HORIZONTAL ANGLES

All movements are made with the upper plate clamp and slow motion screw only. The lower clamp must be locked and neither it, nor its slow motion screw, touched.

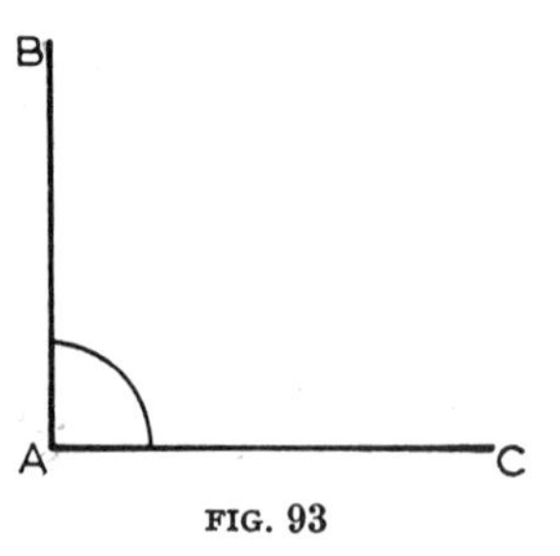

FIG. 93

Only one of the two verniers is read, but it must be the same vernier each time—they are usually lettered A and B. To measure the horizontal angle BAC in fig. 93.

1. Set up the instrument over point A.

2. On face left intersect point B with the vertical hair using the upper plate clamp

and slow motion screw. Read the vernier and book the reading on line B in the face left column.

3. Still on face left intersect C in the same way, read and book the reading of line C in the face left column.

4. On face right intersect C in the same way, read and book the reading on line C in the face right column.

5. On face right intersect B in the same way, read and book on line B in the face right column.

At	*To*	*Face Left*	*Face Right*	*Mean*	*Angle*
A	B	63 25 20	243 25 40	63 25 30	
	C	156 48 40	336 49 00	156 48 50	93 23 20

A set of readings will be obtained as above.

Since the instrument is rotated 180° when changing face the difference between the face left and face right readings is 180°, but not exactly so. In the example it is 180° 00′ 20″, the 20″ being **collimation error.**

The mean is obtained by applying 180° to the face right value and meaning the value obtained with the face left value.

The angle is the difference of the means.

When more than one angle is to be measured (see fig. 94), the procedure is basically the same as before. The reading of the points are taken in a clockwise direction on face left and booked on their respective lines in the face left column. The process is then repeated on face right, reading back in an anti-clockwise direction and booking in the face right column. The only difference is that in this case A, the starting station is read twice on each face to close the circle, this checks that there has been no movement of the instrument during the readings.

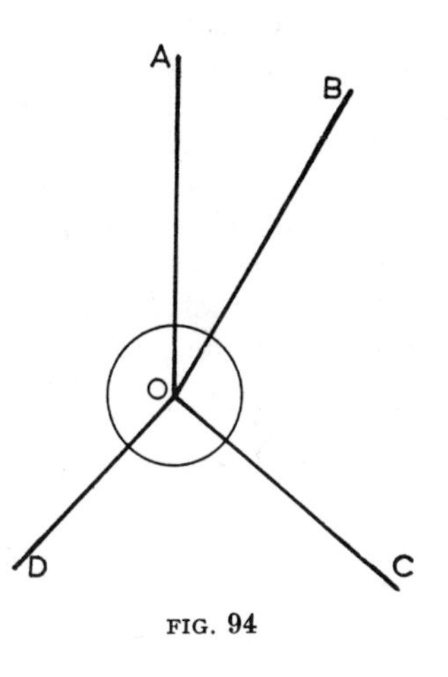

FIG. 94

A set of readings as below will be obtained.

Note that the collimation error is not exactly the same in each case. It would be the same if each pointing was perfect, but errors in setting the cross hairs exactly onto the target each time affect the collimation error.

At	*To*	*Face Left*	*Face Right*	*Mean*	*Angle*
0	A	12 16 00	192 16 40	12 16 20	
	B	43 39 20	223 40 20	43 39 50	31 23 15
	C	141 06 20	321 07 40	141 07 00	97 27 10
	D	207 53 40	27 54 20	207 54 00	66 47 00
	A	12 16 20	192 17 20	12 16 50	164 22 35

Two means are obtained for A, the mean of these means, 12° 16′ 35″, being used when finding the angles.

The above sets of readings represent one **round of angles** in each case. For greater accuracy several rounds can be taken and the mean value for each angle accepted. Each round should be taken with a different reading for A on face left to eliminate errors in graduating the circle.

MEASURING VERTICAL ANGLES

Since the instrument measures the vertical angle (the angle between the horizontal and the line of sight to the station) directly, only one pointing on each face is required and each station is treated independently.

In the example given, it is assumed that the circle is graduated in each direction from 0° at the horizontal. With other types of graduation the vertical angle must be deduced by inspecting the circle and deciding which multiple of 90° represents the horizontal, and then applying the circle readings to this value.

As in the case of the horizontal angles, only one vernier, C or D is read.

To read the vertical angle from point A to point B.

1. Set up instrument at point A.
2. On face left set the horizontal hair on the target.
3. Centre the vertical bubble with the clip screw.
4. Read the vertical vernier and book the value in the face left column.
5. Repeat the process on face right booking the value in the face right column.

At	*To*	*Face Left*	*Face Right*	*Mean*	*Angle*
A	B	03 56 20	03 57 00		03 56 40

MEASURING HORIZONTAL ANGLES BY REPETITION

This is a method of obtaining angles, normally small ones, to a greater accuracy than by the normal method.

To measure the angle ABC.

Set up the instrument at B, sight onto A on face left and note the circle reading; sight onto C with the upper plate clamp and slow motion screw, sight onto A with the lower plate clamp and slow motion screw, sight onto C with the upper plate clamp and slow motion screw, carry on in this way keeping a note of the number of times C has been sighted to, finally read the circle when sighted on C.

The angle on face left is the difference between the two circle readings, divided by the number of repetitions, the number of times C was sighted.

Repeat the process on face right and the accurate angle is the mean of the two values.

To Range a Straight Line with a Theodolite

To prolong AB (fig. 95) from B:

Set up at B and sight on A on face left.

Transit the telescope (i.e. turn it through 180° vertically) and fix C′.

Turn back on to A on face right, transit as before and fix C″. The line may then be extended through point C midway between C′ and C″.

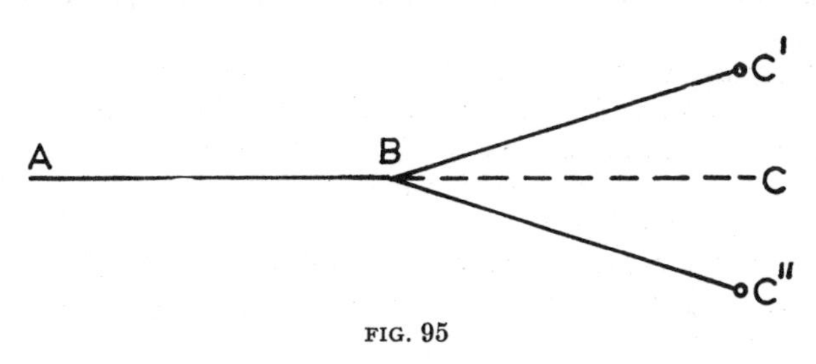

FIG. 95

PERMAMENT ADJUSTMENTS OF THE THEODOLITE

1. Axis of Plate Bubble at Right Angles to the Vertical Axis

This adjustment is similar to the first adjustment of the Dumpy Level.

Test—(*a*) Level the instrument with the foot screws in the normal manner, using the plate bubble.

(*b*) Turn through 180°, and if the bubble moves off centre the instrument is out of adjustment.

Adjustment—(*c*) Take up half the error on the foot screws and the remainder on the bubble adjusting screws.

2. Horizontal Collimation Adjustment

This is to set the line of sight perpendicular to the axis of the trunnions.

Test—On face left intersect a clearly defined target and read the horizontal circle. Repeat on face right (note that this is done every time a horizontal angle is read). If the instrument is in adjustment the readings should be 180° different.

Adjustment—Obtain the mean reading by subtracting 180° from the face right value and meaning this with the face left value. On face left set this mean reading on the circle with the upper plate slow motion screw. The hair will now be off the target. Set it onto the target by moving the diaphragm laterally with its adjusting screws (see page 24). Finally tighten the screws and repeat the test to check.

3. Vertical Hair Adjustment

To set the vertical hair perpendicular to the trunnion axis. This adjustment should always be checked after carrying out the horizontal collimation adjustment as the diaphragm can twist as well as move laterally.

Test—On either face set the vertical hair on a convenient mark. Turn the vertical slow motion screw so that the mark moves the entire length of the hair. If it remains on the hair the instrument is in adjustment. If it moves off the hair carry out the adjustment.

Adjustment—Slightly loosen all the diaphragm adjusting screws and twist it to bring the hair vertical by trial and error. Finally tighten the screws and repeat the test to check.

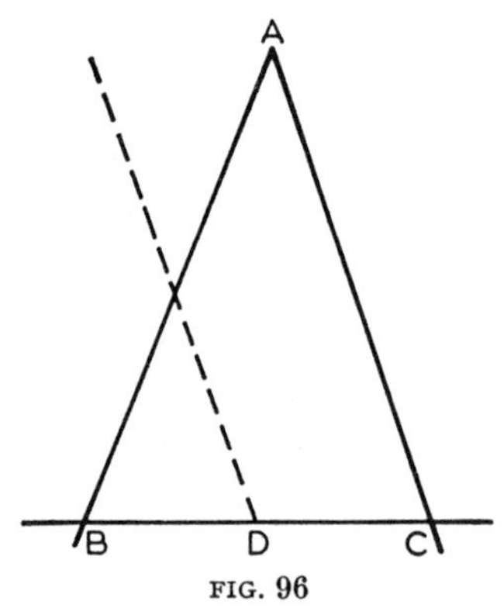

FIG. 96

4. Trunnion Axis Adjustment

To set the trunnion axis perpendicular to the vertical axis. It will then be horizontal when the instrument is levelled.

Test—Set up the instrument near a tall building and on face left intersect a point A at the top. Depress the telescope and mark the point B where the line of sight hits the ground. Repeat on face right marking C. If B and C are the same point the instrument is in adjustment. See fig. 96.

Adjustment—Select D midway between B and C. Intersect D and elevate the telescope. It will miss A. Intersect A by raising or lowering one end of the trunnion axis with the adjusting screws provided. Tighten the screws and repeat the test to check.

5. Vertical Collimation Adjustment

To ensure that the readings on each face are 0° (or a multiple of 90° according to the circle graduations) when the line of sight is horizontal.

Test—On face left intersect a clearly defined target; centre the vertical bubble and read the vertical circle. Repeat on face right. (Note that this is done every time a vertical angle is read.) If instrument is in adjustment, the angles should be the same (or their difference from a multiple of 90° should be the same).

Adjustment—Obtain the circle reading for the mean of the two vertical angles. Set the horizontal hair on the mark with the vertical slow motion screw. Set the circle reading on the vernier with the clip screw. The vertical bubble will not be central, so centralise it with the bubble adjusting screw. Tighten the screws and repeat the test to check.

SOURCES OF ERROR

ERRORS IN USE

1. **Inaccurate Centring**—The hook supporting the plumb line may be bent off centre.

2. **Inaccurate Bisecting of Stations**—(*a*) When a peg forms the permanent station and a ranging rod placed behind the peg forms the temporary station, it must be exactly behind the peg in the line of sight of the theodolite. A pencil may be used in lieu of a ranging rod or a plumb bob string.

(*b*) If a ranging rod is used, it must be held truly vertical, and to avoid error it should be bisected by the cross hairs, as near to the foot as possible.

3. **Parallax**—When the cross hairs and image move in relation to each other when the eye is moved. To correct see page 27.

4. If the vertical axis of the instrument is not truly vertical, angles between the stations, where a considerable difference of level exists, cannot be true. Eliminate by taking face-left and face-right readings.

5. If the horizontal axis is not truly at right angles to the vertical axis, vertical angles will not be measured in a truly vertical plane, and horizontal angles between stations of differing levels will not be accurate. This may be eliminated by taking face-right and face-left readings.

6. **Line of Collimation not at Right Angles to Horizontal Axis**—Eliminate by taking face-left and face-right readings.

7. **Line of Collimation not in centre of Telescope**—This will cause errors in reading of horizontal angles. Eliminate by taking face-left and face-right readings.

8. **Imperfect Graduations of Circles**—Partially remedied by repeated readings using different parts of the circle.

9. **Slip on Bearings when Using Slow-motion Screws**—Come

onto the target with a clockwise movement of the screw.

10. Eccentricity of Plate Bearings—Eliminate by taking face-left and face-right readings.

11. Personal Errors

(*a*) Observing the wrong station.
(*b*) Reading the angle incorrectly.
(*c*) Reading the wrong vernier.
(*d*) Using the wrong slow-motion screw.

Example

What three instrumental errors in a theodolite are eliminated by averaging the 'face-right' and 'face-left' readings of the circles. How would you detect and adjust the errors?

Example

Describe with the aid of sketches and numerical example, where applicable, how you would test a transit theodolite for (*a*) horizontal and (*b*) vertical collimation errors.

State the steps you would take to eliminate any error or apply any correction if you were using an unadjusted instrument for:

(i) setting out a right angle;
(ii) prolonging a straight line;
(iii) measuring the horizontal angle from station A between lines AB and AC, if AB is horizontal and the slope along AC is about 40°;
(iv) measuring the difference of level between B and C.

6 TRAVERSE SURVEYING

DEFINITION

A TRAVERSE is a continuous framework of lines connecting a number of points, the lengths of the lines and their angular relationship to each other being measured. The lines are known as legs and the points as stations.

USE

Traverse surveys are used where site conditions make the chain triangulation method impossible, i.e. a wood, built-up factor block, long winding river, or where the survey is of a large area and details are not required.

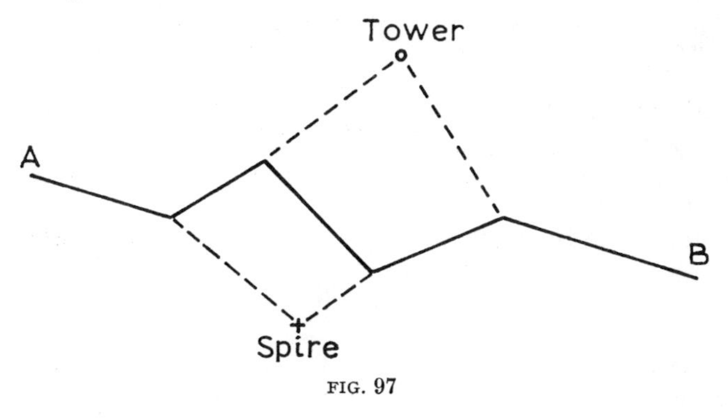

FIG. 97

TYPES

(1) CLOSED TRAVERSE—When the framework forms a closed figure (or when the traverse connects two stations whose positions are known), it is known as a 'closed traverse'. This type of traverse is used for surveying woods, lakes, building blocks or other areas across which no ties, or check lines, can be run. Such a traverse is easily checked, as the survey starts and finishes at a fixed point or points.

(2) OPEN TRAVERSE (see fig. 97)—A traverse whose starting and finishing stations do not coincide or are not both fixed or known points. This type of traverse is used to survey rivers, roads or railway routes. To enable the work to be checked, sights are taken on to some reference object, such as a church tower, factory chimney or other well-defined landmark.

METHODS

(1) Chain.
(2) Chain and compass.
(3) Theodolite.

CHAIN TRAVERSE (see page 12).

COMPASS TRAVERSE (see fig. 98)—The prismatic compass (see page 83) always points towards magnetic north, so that when making a

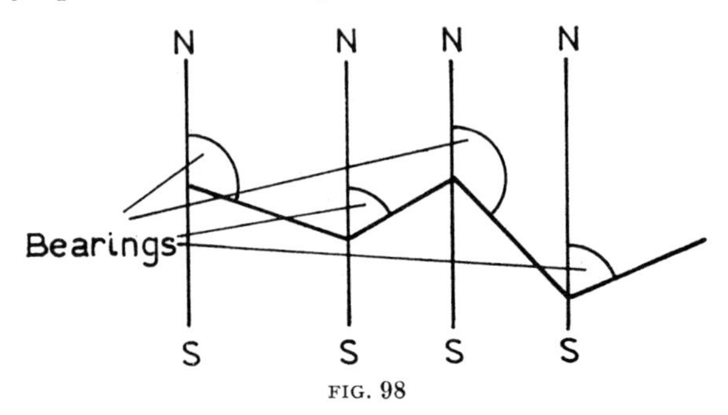

FIG. 98

compass traverse the angles of the legs are related to a north–south line or meridian.

Bearing—The term bearing refers to the angle between the line and the north–south meridian.

Whole-circle Bearing (see fig. 99)—The bearing from north to the

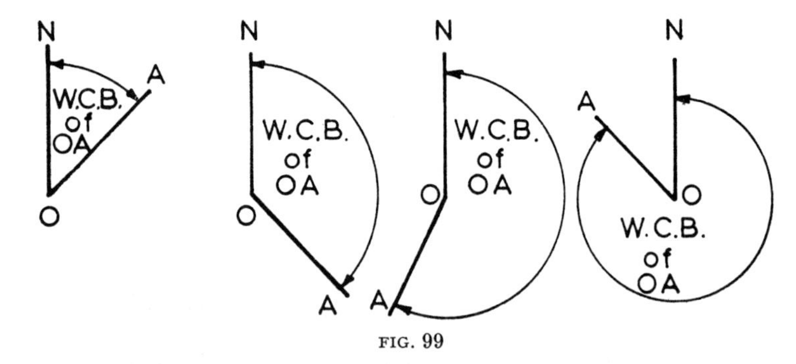

FIG. 99

leg is measured in a clockwise direction and the angle from the north line right round to the leg is known as a whole-circle bearing (abbreviated to W.C.B.).

Forward and Back Bearings—With A as the starting station of the traverse (see fig. 100) and the direction of the survey being towards B, the angle between north and AB at A is known as the forward bearing of AB. The angle between north and AB at station B is known as the back bearing of AB, this should differ from the forward bearing by exactly 180°.

Forward bearing AB = 60°.
Back bearing BA = 240°.
Forward bearing BC = 100°.
Back bearing CB = 280°.

The forward and back bearing of a leg will differ by 180° except where there is 'local attraction'. The presence of metal, metallic ores or electric currents will divert the compass needle from the north–south line, and thus cause the readings taken to be inaccurate. Stations should therefore be chosen so that they are beyond the influence of this attraction.

Correction for Effects of Local Attraction (see fig. 101)

Line	*Bearing*	*Difference of Forward and Back Bearings*
AB	60°	180°
BA	240°	
BC	120°	180°
CB	300°	
CD	210°	178°
DC	32°	
DA	317°	182°
AD	135°	

Since the forward and back bearings at stations A, B and C differ by 180°, these stations must be free from local attraction. The forward and back bearings at D, however, do not vary by this amount, and there must therefore be local attraction at D. This attraction causes the difference of the bearings to be 182° instead of 180°, so that if 2° is deducted from each of the bearings at D (known as 'applying a correction of —2°), the bearings at D will be correct.

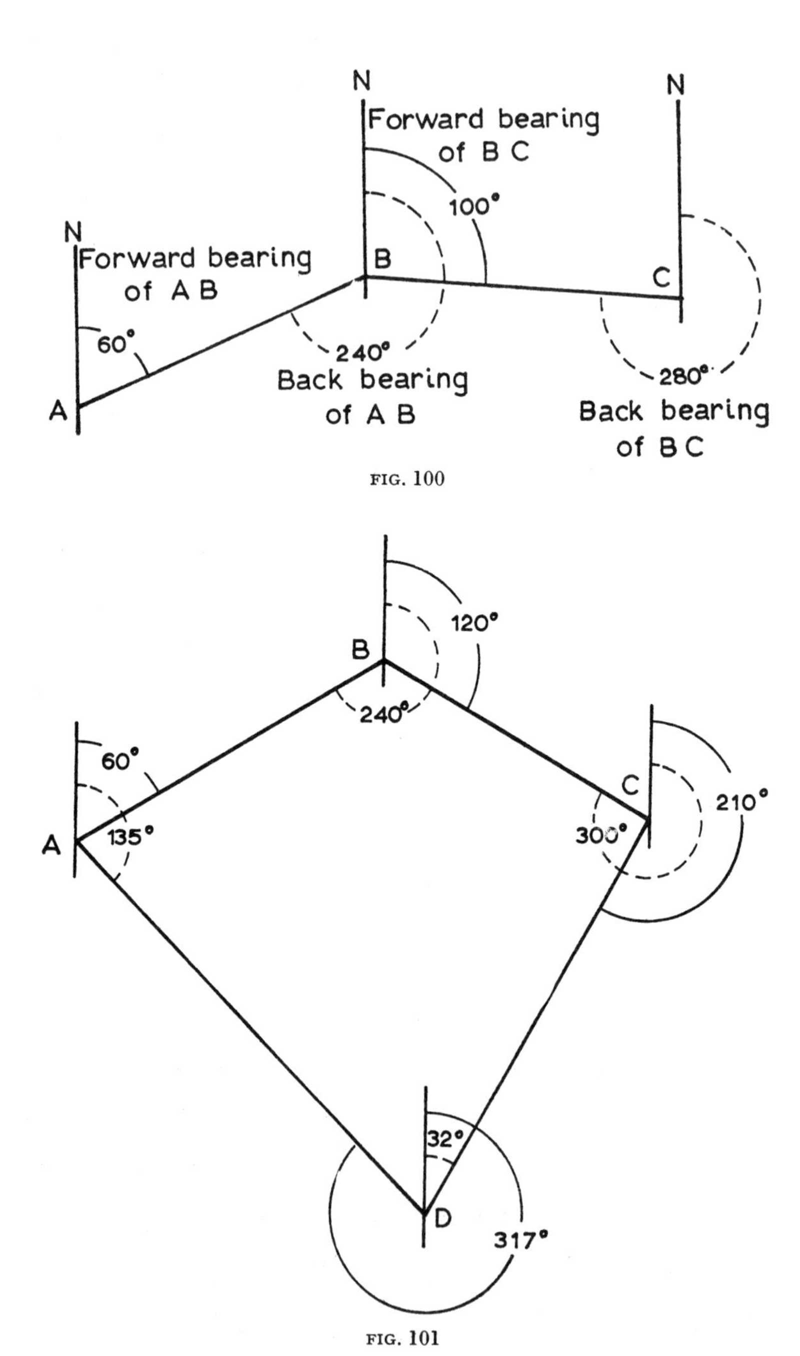

FIG. 100

FIG. 101

Line	*Bearing*	*Difference of Forward and Back Bearing*	*Correction*	*Corrected Bearing*
AB	60°	180°	0	60°
BA	240°		0	240°
BC	120°	180°	0	120°
CB	300°		0	300°
CD	210°	178°	0	210°
DC	32°		−2°	30°
DA	317°	182°	−2°	315°
AD	135°		0	135°

Example

Line	*Observed Bearing*	*Correction*	*Corrected Bearing*
AB	$72\frac{3}{4}°$		
BA	252°		
BC	349°		
CB	$167\frac{1}{4}°$		
CD	$298\frac{1}{2}°$		
DC	$118\frac{1}{2}°$		
DE	229°		
ED	48°		
EA	$135\frac{1}{2}°$		
AE	319°		

To solve this example, first find two adjoining stations (i.e. each at the ends of the same leg) where the forward and back bearings differ by 180°. These stations are not affected by local attraction, and all bearings taken from them will be correct. C and D comply with this requirement. Thus bearings CB, CD, DC and DE all have a correction of 0°. The difference between BC and CB is $349° - 167\frac{1}{4}° = 181\frac{3}{4}°$, therefore station B has a local attraction of $+1\frac{3}{4}°$, and $1\frac{3}{4}°$ must be subtracted from all bearings taken at B, making the correct reading BC $= 347\frac{1}{4}°$ and BA $= 250\frac{1}{4}°$. The difference between AB and BA $= (72\frac{3}{4}° + 360°) - 250\frac{1}{4}° = 182\frac{1}{2}°$. Therefore station A has a local attraction of $+2\frac{1}{2}°$, and $2\frac{1}{2}°$ must be subtracted from all readings at A. The correct bearings being AB $= 70\frac{1}{4}°$ and AE $= 316\frac{1}{2}°$. The completed field book is given below.

Line	*Observed Bearing*	*Correction*	*Corrected Bearing*
AB	$72\frac{3}{4}°$	$-2\frac{1}{2}°$	$70\frac{1}{4}°$
BA	252°	$-1\frac{3}{4}°$	$250\frac{1}{4}°$
BC	349°	$-1\frac{3}{4}°$	$347\frac{1}{4}°$
CB	$167\frac{1}{4}°$	0°	$167\frac{1}{4}°$
CD	$298\frac{1}{2}°$	0°	$298\frac{1}{2}°$
DC	$118\frac{1}{2}°$	0°	$118\frac{1}{2}°$
DE	229°	0°	229°
ED	48°	+1°	49°
EA	$135\frac{1}{2}°$	+1°	$136\frac{1}{2}°$
AE	319°	$-2\frac{1}{2}°$	$316\frac{1}{2}°$

Adjusting Compass Traverse

A compass traverse is plotted with protractor and scale and, since the error is likely to be in the neighbourhood of 1 in 400, such a traverse will never close exactly. The traverse shown in fig. 102 does not close by a distance AA_1.

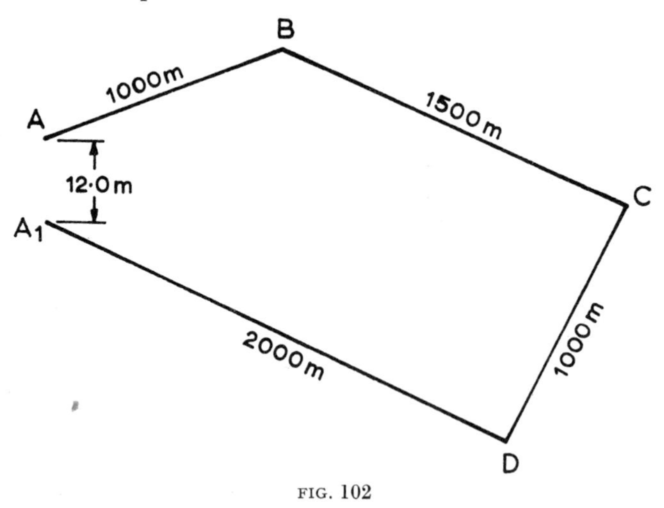

FIG. 102

First determine whether the error AA_1 is within the permissible error of 1 in 400.

The total length of the traverse $= 1000 + 1500 + 1000 + 2000$
$= 5500$ m.

Let distance AA_1 be equal to 12 m.

$$\text{Then error} = \frac{\text{Closing error}}{\text{Length of traverse}} = \frac{12}{5500} = \frac{1}{458}.$$

Though the error 12 m appears to be excessive, there would be no point in repeating the work using a chain and compass since an error of approximately 1 in 400 would still be present due to the limitations of the instruments. The existing survey should therefore be adjusted to make it close. A method commonly used is known as Bowditch's Method, and is applied as follows:

1. Draw a line and set off to scale on it the distances AB, BC, etc. (see fig. 103).

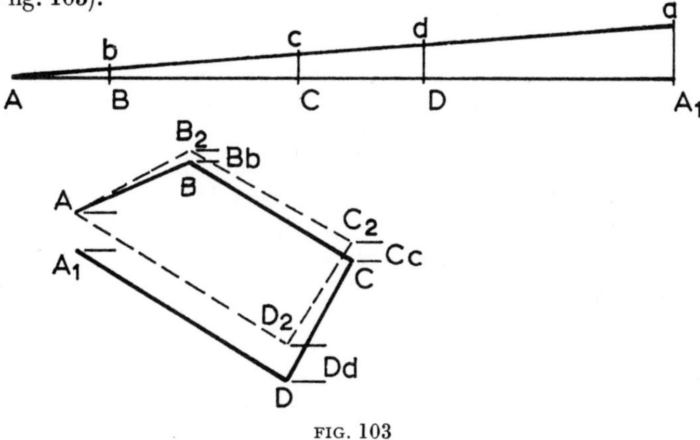

FIG. 103

2. At A_1 on the line draw a line A_1a perpendicular to AA_1 and equal in length to the closing error A_1A on the drawing.
3. Join A*a*.
4. From B, C and D on the line draw lines B*b*, C*c*, D*d* perpendicular to AA_1 to intersect line A*a*.
5. Join AA_1 on the drawing.
6. From B on the drawing set out a line equal in length to B*b* and parallel to AA_1 to point B_2. Fix points C_2 and D_2 in a similar manner.
7. Join A, B_2, C_2, D_2, and the resultant figure will be the adjusted traverse.

Notes on Compass Traverse

This method is not used for accurate work since the compass can normally be read only to about the nearest $\frac{1}{2}°$. It is sometimes used in reconnaissance surveys, where a rough map is required, the lengths of the legs being measured with a chain or paced out.

Advantages—The compass is light and easy to carry, bearings can be taken much more quickly than with the theodolite, and it eliminates the need for check lines and so forms a means of rapid survey.

The bearing of each leg is determined independently so that any error in the bearing can be confined to that particular leg.

Disadvantages—The degree of accuracy is not high, and the instrument readings can be affected by 'local attraction'.

Theodolite Traverses

For description and method of use of the theodolite see page 85.

USE

Where a high degree of accuracy is required.

METHOD

1. **Fix the Stations**—Stations, which are located by wooden pegs, should be as few as possible in number. They should be visible from the preceding and following stations, and should be 'tied in' to fixed points, such as corners of buildings, kerbs, etc., in order that they can be easily relocated should the pegs be destroyed. When working in towns the stations should be marked by driving nails between the paving-stones or by a cross being marked on the stone.

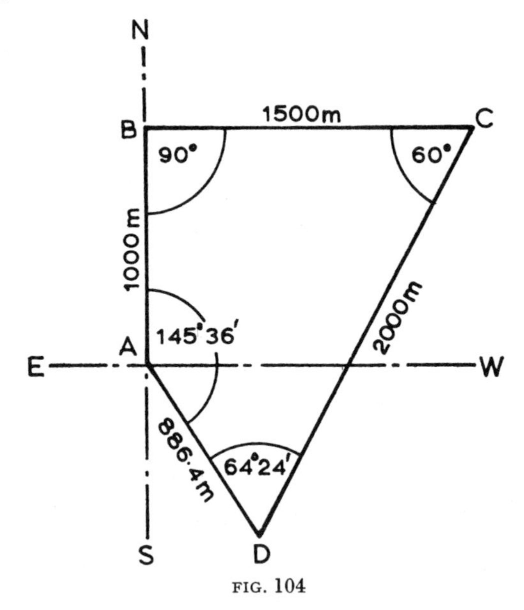

FIG. 104

2. **Set up at A** (see fig. 104)—If a compass is attached to the theodolite, the bearing of the line AB may be taken This line is called the meridian of the traverse. If there is no compass attached to the theodolite (and in most modern instruments this is the case), an arbitrary meridian may be established by setting up the instrument over station A and directing it on to a permanent landmark and calling this the

north–south line. The true bearing of the arbitrary meridian can be determined later with a compass or measured from an Ordnance map.

3. **Measure the Angle BAD**—Since the horizontal circle is graduated clockwise, it is better to measure the angles in a clockwise direction, otherwise the readings will have to be subtracted from 360° (or 180°, according to which vernier is read) (see fig. 105). See page 89.

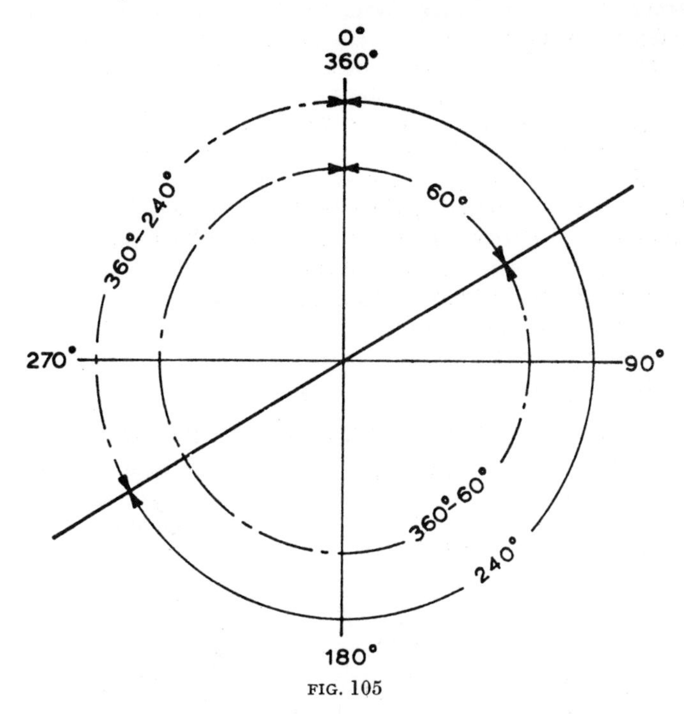

FIG. 105

4. **Measure the Line AB**—The lengths of the legs should be measured with a steel tape graduated in metres and five millimetre units to ensure maximum accuracy. It would be pointless to measure the angles to an accuracy of, say, $\frac{1}{10000}$ and then to use a chain to measure the legs to an accuracy of $\frac{1}{500}$.

5. **Measure the Angle CBA**—Set up at B, sight on to C and read angle CBA. The other sides and angles are measured in a similar manner.

PLOTTING THE TRAVERSE (shown in fig. 104)

1. Draw a line east–west across the paper.

2. At some convenient point on the line draw a north–south line. Let the intersection of the lines represent the position of A.

3. To plot the position of B in the example given, set out 300 m to scale on the north–south line and fix the position of B.

4. From B, set out at right angles to the north–south line 450 m to scale and fix the position of C.

5. Line CD could be set out using a protractor to make an angle of 60° with line BC, 600 m being set out from C to fixed point D. **However, a protractor, even a large one, only measuring to a quarter of a degree, and as the angles have been measured with considerable accuracy, a more accurate method of fixing the stations is needed. This is done by calculating the 'co-ordinates' of the stations.**

Co-ordinates

To fix the co-ordinates involves the following steps:

1. Check and correct the angles at the stations (internal or external).
2. Convert the angles to **reduced bearings.**
3. Calculate the **latitudes and departures.**
4. Calculate the **co-ordinates.**
5. Correct the co-ordinates.

Each of the above steps is dealt with as follows:

1. Check and Correct the Angles at the Stations.

The sum of the internal angles of any closed figure should equal $(2n - 4)$ right angles, where n = number of sides of the figure.

The sum of the external angles of any figure should equal $(2n + 4)$ right angles.

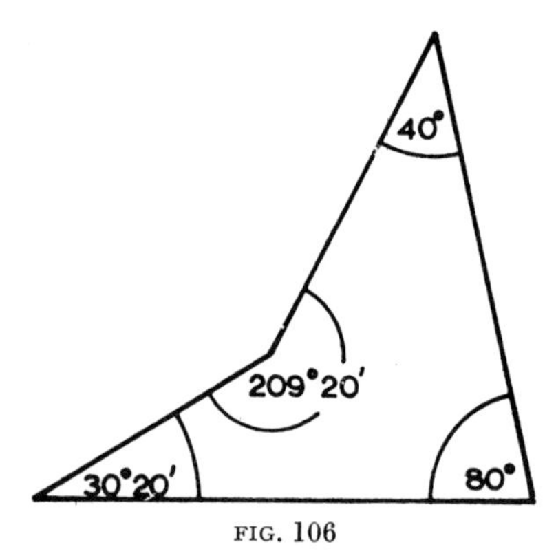

FIG. 106

Checking the traverse shown in fig. 104.

$$\text{Sum of the internal angles} = (2n - 4) \times 90°$$
$$90° + 60° + 64°\,24' + 145°\,36' = (2 \times 4 - 4) \times 90°$$
$$360° = 360°.$$

The internal angles in this traverse have been measured correctly. Consider now the traverse shown in fig. 106.

$$\text{Sum of the internal angles} = 40° + 80° + 30° 20' + 209° 20'$$
$$= 359° 40'.$$

This is an error of $-20'$.

This is a considerable error for a theodolite traverse, and the angles should be reread on the site. However, in order to demonstrate the method of correcting the error assume that this error is permissble. To make the traverse close angularly, the error is divided amongst **all** the angles; that is, each angle is increased by $\frac{1}{4} \times 20'$.

$$\text{Angle A} = 209° 20' \quad +(\tfrac{1}{4} \times 20')$$
$$= 209° 20' \quad +5'$$
$$= \mathbf{209° 25'}.$$

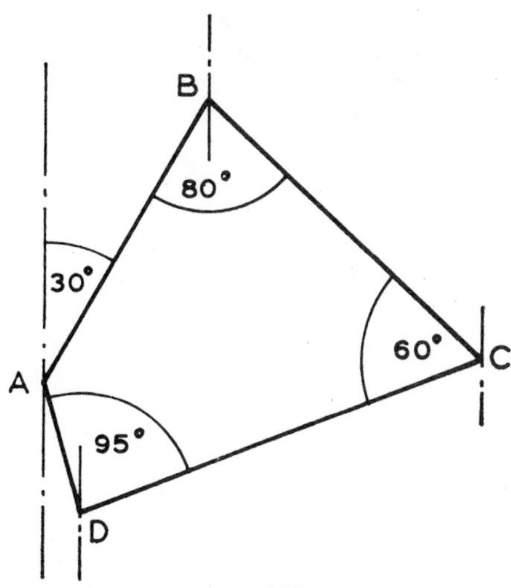

FIG. 107

$$\text{Angle B} = 40° \quad +(\tfrac{1}{4} \times 20')$$
$$= 40° \quad +5'$$
$$= \mathbf{40° 5'}.$$

$$\text{Angle C} = 80° \quad +(\tfrac{1}{4} \times 20')$$
$$= 80° \quad +5'$$
$$= \mathbf{80° 5'}.$$

$$\text{Angle D} = 30° 20' \quad +(\tfrac{1}{4} \times 20')$$
$$= 30° 20' \quad +5'$$
$$= \mathbf{30° 25'}.$$

$$\text{Sum of the corrected angles} = 209° 25' + 40° 5' + 80° 5' + 30° 25'$$
$$= \mathbf{360°}.$$

2. Converting the Angles to Reduced Bearings

The reduced bearing (R.B.) of the leg of a traverse is the **smallest** angle it makes with the north–south meridian (see fig. 107).

It is best to calculate the W.C.B. of the legs first.

The W.C.B. of AB = 30°.

To find the W.C.B. of BC consider fig. 108.

Angle NBA′ = W.C.B. AB = 30°.
Angle A′BC = 180° − Int. angle B = 100°.
Angle NBC = W.C.B. BC
= 30° + (180° − 80°)
= 130°.

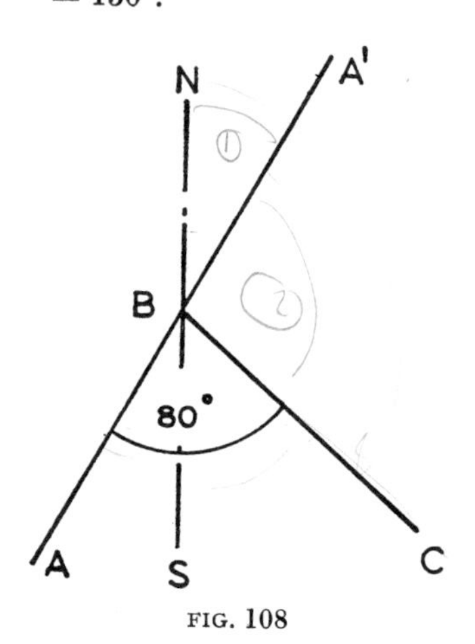

FIG. 108

To find the W.C.B. of CD consider fig. 109.

Angle NCB′ = W.C.B. BC = 130°.
Angle B′CD = 180° − Int. angle C = 120°.

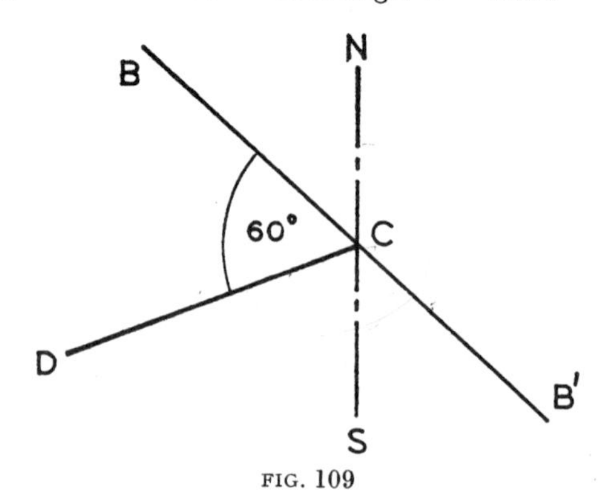

FIG. 109

Angle NCD = W.C.B. CD
= W.C.B. BC + (180° − Int. angle C)
= 130° + (180° − 60°)
= 250°.

To find the W.C.B. of DA, consider fig. 110.

Angle NDC′ = W.C.B. CD = 250°.
Angle C′DA = 180° − Int. angle D = 85°.
Angle NDA = W.C.B. DA
= W.C.B. CD + (180° − Int. angle D)
= 250° + (180° − 95°)
= 335°.

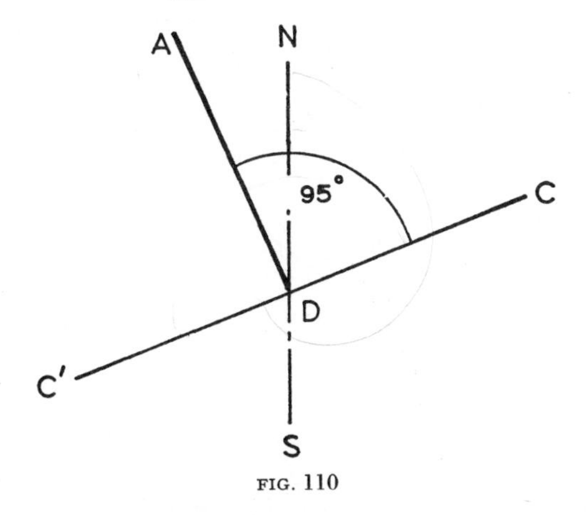

FIG. 110

It will be seen that the W.C.B. of a leg is equal to the W.C.B. of the previous leg plus 180° minus the internal angle at the station. As a check calculate the W.C.B. of AB.

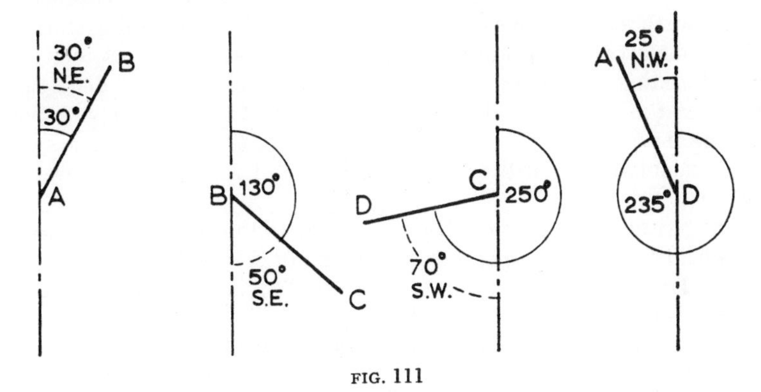

FIG. 111

W.C.B. AB = W.C.B. DA + (180° − Int. angle A)
= 335° + 180° − 125°
= 390° or 30° which agrees with the given W.C.B.

The W.C.B. are now converted to R.B. (see fig. 111).

W.C.B. AB = 30°; W.C.B. BC = 130°; W.C.B. CD = 250°; W.C.B. DA = 335°; R.B. AB = 30° NE; R.B. BC = 50° SE; R.B. CD = 70°; R.B. DA = 25° NW.

3. Calculating the Latitudes and Departures (Northings and Southings° Eastings and Westings)

The latitude of a station is its distance north or south, sometimes called northing or southing, of the preceding station; positive latitude

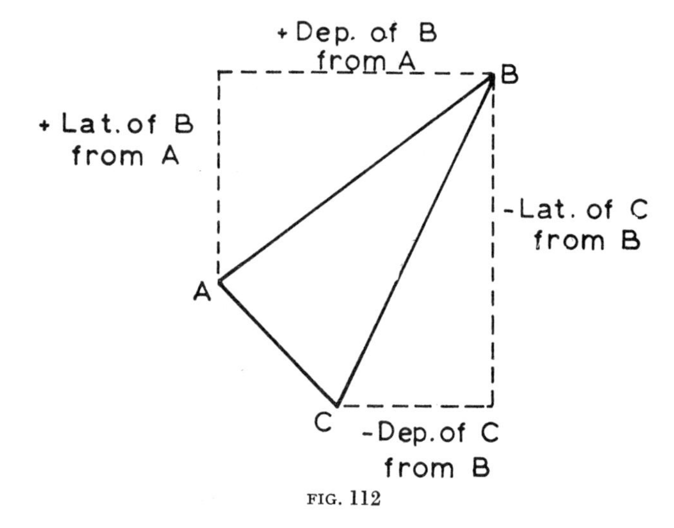

FIG. 112

being north and negative latitude being south of the preceding station.

The departure of a station is its distance east or west, sometimes called easting or westing, of the preceding station; positive departure being east and negative departure being west of the preceding station (see fig. 112).

To calculate the latitudes and departures of the traverse shown in fig. 104.

The latitude of B (see fig. 113) = +1000 and the departure of B = 0.

The latitude of C = 0 and the departure of C = +1500.

The latitude of D (see fig. 114) = −1732·2 and the departure of D = −1000.

The latitude of A relative to D = +732·2 (see fig. 115).

The departure of A relative to D = −500.

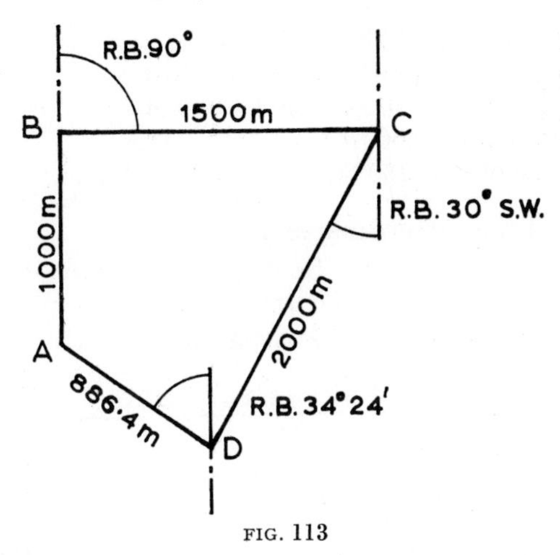

FIG. 113

It will be noted that to calculate the latitudes of D and A, the length of the leg is multiplied by the cosine of the reduced bearing. Thus we get the rule 'the latitude of a station is the length of the leg × cos reduced bearing of the preceding station' *or* **latitude = length × cos R.B.** Similarly the **departure = length × sin R.B.**

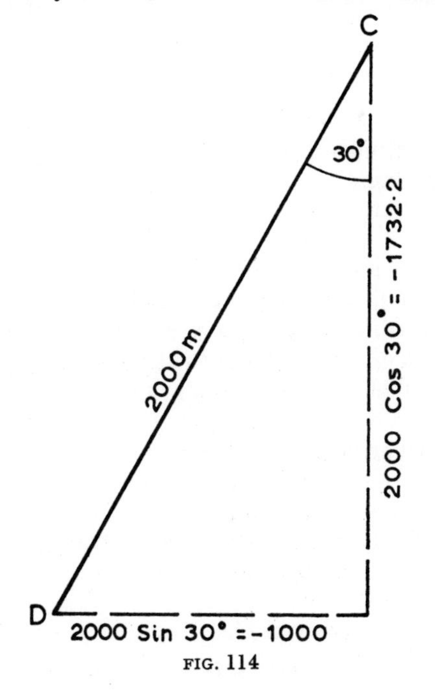

FIG. 114

It has been stated that the latitude of B was 1000 m. This should equal length $\times$ cos R.B $= 1000 \times \cos 0° = 1000$ m.

The departure of B $= 1000 \times \sin 0° = 0$ m.

The latitude of C $= 1500$ m $\times \cos 90° = 0$ m.

The departure of C $= 1500$ m $\times \sin 90° = 1500$ m.

4. Calculating the Co-ordinates

The co-ordinates of a station fix its position relative to some origin, usually the first station of the traverse. The co-ordinates are the algebraic sum of the latitudes and departures respectively.

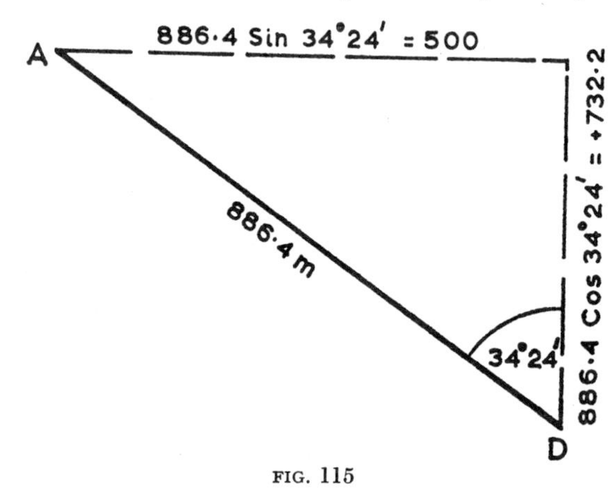

FIG. 115

In the case of latitudes, if the sum is positive it is called a northing, if negative a southing.

Similarly, departures are called eastings if positive and westings if negative.

If A is the origin, then its co-ordinates are 0 m N. and 0 m E.

The co-ordinates of B are 1000 m N. and 0 m E.

The co-ordinates of C are 1000 m N. and 1500 m E.

The co-ordinates of D are Sum of the latitudes

$$= (+1000 - 1732{\cdot}2) \text{ m}$$
$$= \mathbf{-732{\cdot}2 \text{ m}}$$

and Sum of the departures

$$= (+1500 - 1000) \text{ m}$$
$$= \mathbf{+500 \text{ m.}}$$

The co-ordinates of A $= (1000 - 1732{\cdot}2 + 732{\cdot}2)$ m

$$= \mathbf{0 \text{ m N.}}$$

and $(1500 - 1000 - 500)$ m

$$= \mathbf{0 \text{ m N.}}$$

A systematic method of booking the results is shown in fig. **116**.

Sta.	*Line*	*Length (metres)*	*W.C.B.*	*R.B.*	*Latitude*	*Departure*	*Co-ordinates*			
							N.	*S.*	*E.*	*W.*
A							0	0	0	0
	AB	1000	0°	N. 0° E.	+1000	0				
B							1000		0	
	BC	1500	90°	N. 90° E.	0	+1500				
C							1000		1500	
	CD	2000	210°	S. 30° W.	−1732.2	−1000				
D								732·2	500	
	DA	886·4	325° 36′	N. 34° 24′ W.	+732·2	−500				
A							0	0	0	0

FIG. 116

It will be seen that when the co-ordinates of A were calculated, they gave the result we started with, i.e. 0 m N. and 0 m E. In practice this is not likely to occur, for the following errors will generally arise even in the most carefully surveyed work.

1. Small errors in measuring lengths (tapes are graduated in units of five millimetres only).

2. Small errors in measuring angles (most theodolites are graduated to 20 secs. only).

3. Calculations should be carried out on five-figure logarithm tables, and even these do not give complete accuracy. In fact, there will always be an error, but, providing that the error does not exceed that permissible (usually $\frac{1}{5000}$ for site surveys), the traverse may be adjusted so that it closes, i.e. the error being distributed proportionally between the co-ordinates.

CORRECTION OF CLOSING ERROR

Station	*Leg*	*Length*	*Latitude*	*Departure*	*N.*	*S.*	*E.*	*W.*
A	AB	496	+297·5	+395	0		0	
B	BC	498	−302·5	+395	297·5		395	
C	CD	504	−302·5	−405		5	790	
D	DA	502	+297·5	−405		307·5	385	
A						10		20
Total length, 2000 m.					Error	10 S.		20 W.

The above traverse does not close and the proportional error is

$$\frac{\text{Closing error}}{\text{Total length of traverse}}.$$

Therefore in latitude the error is $\frac{10}{2000} = \frac{1}{200}$ and in departure it is $\frac{20}{2000} = \frac{1}{100}$. These are considerable errors for a theodolite traverse.

However, in order to demonstrate the method of correcting the error, assume that these errors are permissible. To close the traverse apply to each latitude an amount equal to length of line $\times \frac{1}{200}$ and to each departure an amount equal to length of line $\times \frac{1}{100}$. This is known as Bowditch's Method.

$$\begin{aligned}
\text{Corrected latitude of B} &= +297{\cdot}5 + (496 \times \tfrac{1}{200}) \\
&= +297{\cdot}5 + 2{\cdot}48 \\
&= \mathbf{+300.} \\
\text{Corrected latitude of C} &= -302{\cdot}5 + (498 \times \tfrac{1}{200}) \\
&= -302{\cdot}5 + 2{\cdot}49 \\
&= \mathbf{-300.} \\
\text{Corrected latitude of D} &= -302{\cdot}5 + (504 \times \tfrac{1}{200}) \\
&= -302{\cdot}5 + 2{\cdot}52 \\
&= \mathbf{-300.} \\
\text{Corrected latitude of A} &= +297{\cdot}5 + (502 \times \tfrac{1}{200}) \\
&= +297{\cdot}5 + 2{\cdot}51 \\
&= \mathbf{+300.} \\
\text{Corrected departure of B} &= +395 + (496 \times \tfrac{1}{100}) \\
&= +395 + 4{\cdot}96 \\
&= \mathbf{+400.} \\
\text{Corrected departure of C} &= +395 + (498 \times \tfrac{1}{100}) \\
&= +395 + 4{\cdot}98 \\
&= \mathbf{+400.} \\
\text{Corrected departure of D} &= -405 + (504 \times \tfrac{1}{100}) \\
&= -405 + 5{\cdot}04 \\
&= \mathbf{-400.} \\
\text{Corrected departure of A} &= -405 + (502 \times \tfrac{1}{100}) \\
&= -405 + 5{\cdot}02 \\
&= \mathbf{-400.}
\end{aligned}$$

CORRECTED CO-ORDINATES TABLE

Station	*Corrected Latitude*	*Corrected Departure*	*Corrected Co-ordinates*			
			N.	*S.*	*E.*	*W.*
A	+300	+400	0	0	0	0
B	−300	+400	300		400	
C	−300	−400		0	800	
D	+300	−400		300	400	
A			0	0	0	0

The traverse will now close.

CALCULATION OF THE AREA OF A TRAVERSE

The area of a traverse can be determined without plotting the traverse, provided the co-ordinates of the stations are known, and the result will be more accurate than any actual measurement of the plotted figure would be. Consider the traverse shown in fig. 117.

Let x_1, x_2, x_3, x_4, be the eastings of the stations
and y_1, y_2, y_3, y_4, be the northings of the stations.

Area ABCD
= (Area ABFE + Area BCGF) − (Area CGHD + Area DHEA).
See fig. 118.

$$\text{Area ABCD} = \frac{(y_1 + y_2)}{2}(x_2 - x_1) + \frac{(y_2 + y_3)}{2}(x_3 - x_2)$$

$$- \frac{(y_3 + y_4)}{2}(x_3 - x_4) - \frac{(y_4 + y_1)}{2}(x_4 - x_1).$$

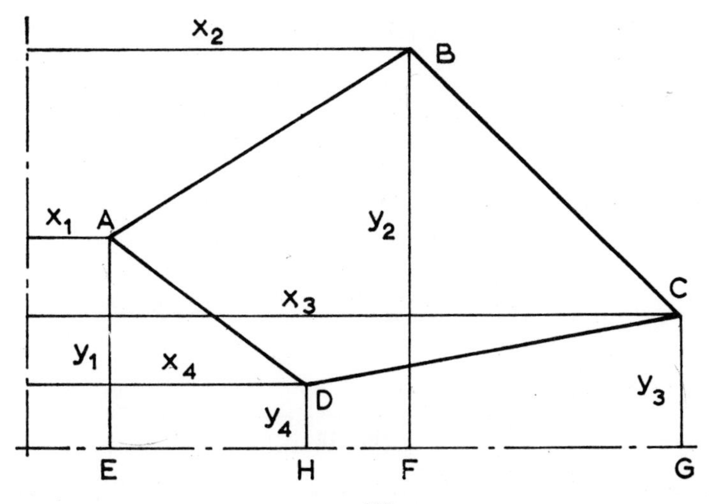

FIG. 117

FIG. 118

Taking the common factor, $\frac{1}{2}$, outside and expanding the expression:

Area ABCD

$$= \tfrac{1}{2}(y_1x_2 - y_1x_1 + y_2x_2 - y_2x_1 + y_2x_3 - y_2x_2 + y_3x_3 - y_3x_2 - y_3x_3 + y_3x_4 - y_4x_3 + y_4x_4 - y_4x_4 + y_4x_1 - y_1x_4 + y_1x_1)$$

$$= \tfrac{1}{2}(x_1y_4 - x_1y_2 + x_2y_1 - x_2y_3 + x_3y_2 - x_3y_4 + x_4y_3 - x_4y_1)$$

$$= \tfrac{1}{2}[x_1(y_4 - y_2) + x_2(y_1 - y_3) + x_3(y_2 - y_4) + x_4(y_3 - y_1)].$$

It will be noticed that each term is the easting of the station × (northing of the preceding station − northing of the following station).

To check the validity of this method, let us take a square of side 4 units and place it in various positions about the axes (see fig. 119).

Area = $\frac{1}{2}$[E.ing of A(N.ing of D − N.ing of B) + E.ing of B (N.ing of A − N.ing of C), etc.]

$= \frac{1}{2}[2(2 - 6) + 2(2 - 6) + 6(6 - 2) + 6(6 - 2)]$

$= \frac{1}{2}(-8 - 8 + 24 + 24)$

$= \frac{1}{2} \times 32$

= 16 square units (which is the area of a square of side 4 units).

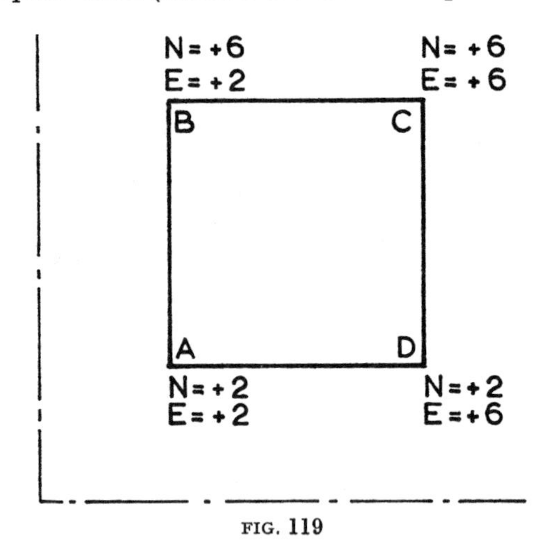

FIG. 119

Fig. 120 is the square of side of 4 units placed centrally round the origin.

Area $= \frac{1}{2}[-2(-2 - (+)2) - 2(-2 - (+)2) + 2(2 - (-)2) + 2(2 - (-)2)]$

$= \frac{1}{2}[(-2 \times -4) + (-2 \times -4) + (2 \times +4) + (2 \times +4)]$

$= \frac{1}{2}(+8 + 8 + 8 + 8)$

$= \frac{1}{2} \times 32$

= 16 square units.

The student is advised to work through the following examples and those at the end of the chapter.

The work involved in calculating traverse surveys may seem complicated and laborious, but after some practice the procedure will become fairly simple.

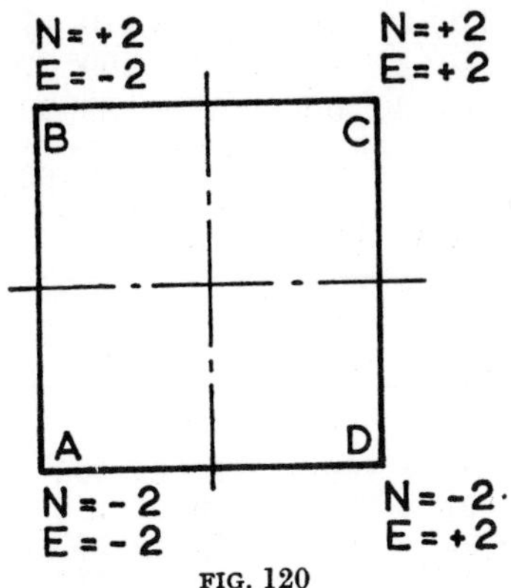

FIG. 120

EXAMPLE OF COMPLETE TRAVERSE CALCULATIONS

Given traverse:

	Int. Angle	*Line*	*Length*
A	126° 1′ 40″	AB	692 m
B	95° 2′ 20″	BC	1,137 m
C	129° 57′ 00″	CD	809 m
D	99° 27′ 00″	DE	1,170 m
E	89° 30′ 40″	EA	1,340 m

1. Check and Correct the Angles

Sum of internal angles = 126° 01′ 40″ + 95° 02′ 20″ + 129° 57′ 00″ + 99° 27′ 00″ + 89° 30′ 40″
= 539° 58′ 40″.

Error = 540° − 539° 58′ 40″
= 1′ 20″.

Correction to add to angles:

$$\frac{1' \, 20''}{5} = 16''.$$

Angle		*Correction*
A	126° 01′ 40″	+16″
B	95° 02′ 20″	+16°
C	29° 57′ 00″	+16″
D	99° 27′ 00″	+16″
E	89° 30′ 40″	+16″
		80″

Corrected angles:

A	126° 01′ 40″ + 16″	=	126° 01′ 56″
B	95° 02′ 20″ + 16″	=	95° 02′ 36″
C	129° 57′ 00″ + 16″	=	129° 57′ 16″
D	99° 27′ 00″ + 16″	=	99° 27′ 16″
E	89° 30′ 40″ + 16″	=	89° 30′ 56″
			540° 00′ 00″

2. Calculating W.C.B.s

Line		*W.C.B.*
AB	=	30° 00′ 00″ Given
BC	30° 00′ 00″ + (180° − 95° 02′ 36″) =	114° 57′ 24″
CD	114° 57′ 24″ + (180° − 129° 57′ 16″) =	165° 00′ 08″
DE	165° 00′ 08″ + (180° − 99° 27′ 16″) =	245° 32′ 52″
EA	245° 32′ 52″ + (180° − 89° 30′ 56″) =	336° 01′ 56″
Check AB	336° 01′ 56″ + (180° − 126° 01′ 56″) =	390° 00′ 00″ or 30°

3. Calculating Latitudes, Departures and Co-ordinates

Sta.	*Line*	*Length (metres)*	*W.C.B.*	*R.B.*	*Lat.*	*Dep.*	*Co-ordinates*			
							N.	*S.*	*E.*	*W.*
A							0	0	0	0
	AB	692	30° 00′ 00″	N. 30° 00′ 00″ E.	+599·3	+346				
B							599·3		346·0	
	BC	1137	114° 57′ 24″	S. 65° 02′ 36″ E.	−481·2	+1030				
C							118·1		1376·0	
	CD	890	165° 00 08″	S. 14° 59′ 52″ E.	−859·6	+230·3				
D								741·5	1606·3	
	DE	1170	245° 32′ 52″	S. 65° 32′ 52″ W.	−486·1	−1065				
E								1227·6	541·3	
	EA	1340	336° 01′ 56″	N. 23° 58′ 04″ W.	+1226·0	−544·4				
A								1·6		3·1
		5229								

Error in co-ordinates 1·6 south, 3·1 west.

4. Correcting Co-ordinates

Line	*Lat. Correction*	*Dep. Correction*
AB	$\frac{1\cdot6}{5229} \times 692 = 0\cdot21$	$\frac{3\cdot1}{5229} \times 692 = 0\cdot41$
BC	,, $\times 1137 = 0\cdot35$	,, $\times 1137 = 0\cdot67$
CD	,, $\times 890 = 0\cdot27$	,, $\times 890 = 0\cdot53$
DE	,, $\times 1170 = 0\cdot36$	,, $\times 1170 = 0\cdot69$
EA	,, $\times 1340 = 0\cdot41$	,, $\times 1340 = 0\cdot80$
	1·60	3·10

Sta.	*Line*	*Lat. Corr.*	*Dep. Corr.*	*Corrected Latitude*	*Corrected Departure*	*Corrected Co-ordinates*			
						N.	*S.*	*E.*	*W.*
A						0	0	0	0
	AB	0·21	0·41	+599·51	+346·41				
B						599·51		346·41	
	BC	0·35	0·67	−480·85	+1030·67				
C						118·66		1377·08	
	CD	0·27	0·53	−859·33	+230·83				
D							740·67	1607·91	
	DE	0·36	0·69	−485·74	−1064·31				
E							1226·41	543·60	
	EA	0·41	0·80	+1226·41	−543·6				
A						0	0	0	0

5. Determination of Areas

Area of traverse $= \frac{1}{2}$ Easting of Station (Northing of preceding station, Northing of following station), etc.

$$= \tfrac{1}{2}[0(-1226{\cdot}41 - 599{\cdot}51) + 346{\cdot}41(0 - 118{\cdot}66) + 1377{\cdot}08(599{\cdot}51 - (-)740{\cdot}67) + 1607{\cdot}91(118{\cdot}66 - (-)1226{\cdot}41) + 543{\cdot}6(-740{\cdot}67 - 0)]$$

$$= \tfrac{1}{2}[0 - (346{\cdot}41 \times 118{\cdot}66) + (1377{\cdot}08 \times 1340{\cdot}18) + (1607{\cdot}91 \times 1345{\cdot}07) - (543{\cdot}6 \times 740{\cdot}67)]$$

$$= \mathbf{1\ 782\ 187\ m^2}.$$

Example

A straight tunnel is to be driven between two points, A and B, whose co-ordinates are given in the following table.

Point	*Co-ordinates (m)*	
A	0	0
B	+3014	+256
C	+1764	+1398

It is desired to sink a shaft at D, the mid point on AB, but it is impossible to measure directly along AB, so D is to be fixed from a third point, C, which is known.

Calculate (1) Co-ordinates of D.
(2) Length and bearing of CD.
(3) Angle ACD, given that the bearing AC is 38° 24′.

Solution (see fig. 121)

$$\text{Co-ordinates of D,} = +\frac{3014}{2}, +\frac{256}{2}$$

$$= \mathbf{+1507, +128.}$$

Bearing of CD:

$$\tan CDX = \frac{257}{1270}$$

$$= 0{\cdot}2024$$

$$CDX = 11^\circ\ 26'$$

$$\text{Bearing of CD} = 270^\circ - 11^\circ\ 26'$$

$$= \mathbf{258^\circ\ 34'}.$$

$$\text{Length of CD} = \frac{\text{DX}}{\sin \text{CDX}}$$
$$= \frac{257}{\sin 11° 26'}$$
$$= \mathbf{1296{\cdot}5\ m.}$$
$$\text{Angle ACD} = 258° 34' - (38° 34' + 180° 0')$$
$$= 40° 10'.$$

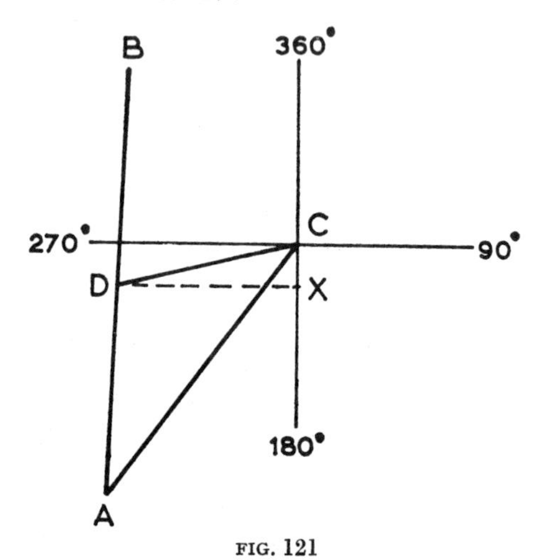

FIG. 121

Example

The co-ordinates of station A are +599·20 m and −255·99 m. A line AX runs at a bearing of 249° 06′. A second station, B, has co-ordinates +508·82 m and −205·48 m. From station B a line BY runs at a bearing of 292° 15′ and meets AX at the point Z. Calculate the distance BZ.

Solution (see fig. 122)

292° 15′ and 249° 06′ are whole-circle bearings.

$$\text{Angle XZY} = \text{Angle AZB} = 292° 15' - 249° 06' = 43° 09'$$
$$\text{Length BQ} = 599{\cdot}20 - 508{\cdot}82 = 90{\cdot}38$$
$$\text{Length AQ} = 255{\cdot}99 - 205{\cdot}48 = 50{\cdot}51$$
$$\therefore \quad \text{Tan QBA} = \frac{50{\cdot}51}{90{\cdot}38} = 29° 12'.$$

Then the bearing of BA = 360° − 29° 12′ = 330° 48′.

$$\therefore \quad \text{Angle ABZ} = 330° 48' - 292° 15' = 38° 33'$$
$$\text{and angle ZAB} = 180° - (43° 09' + 38° 33') = 98° 18'.$$

$$\text{Now length AB} = \frac{50{\cdot}51'}{\sin 29° 12'} = 103{\cdot}53\ \text{m.}$$

$$\text{Therefore } \frac{\text{BZ}}{\sin 98° 18'} = \frac{103{\cdot}53}{\sin 43° 09'}$$

$$\text{and BZ} = \frac{103{\cdot}53 \times \sin 98° 18'}{\sin 43° 09'} = \mathbf{149{\cdot}8\ m.}$$

Example

A tunnel is to be driven from two shafts situated at A and D. The points A and D are linked by the surface traverse (shown in fig. 123).

The line AB has a bearing due eastward. Set out the particulars of the traverse in tabular form and complete the table. Find the length and bearing of the line AD.

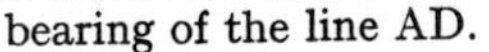

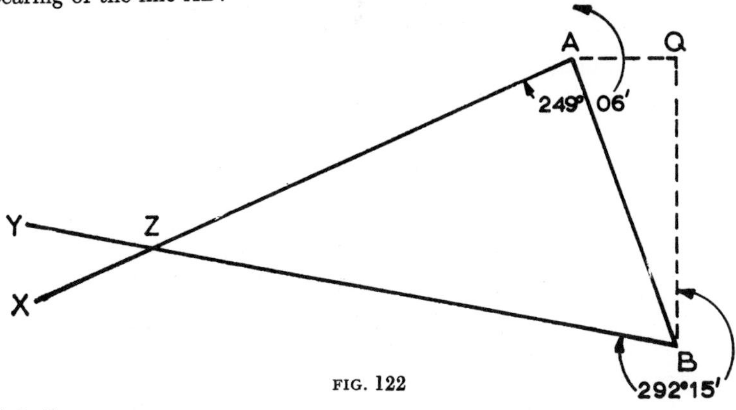

FIG. 122

Solution

Sta.	*Line*	*Length (metres)*	*W.C.B.*	*R.B.*	*Lat.*	*Dep.*	*N.*	*S.*	*E*	*W.*
A										
	AB	1200	90°	N. 90° E.	0	+1200				
B									1200	
	BC	1061	159°	S. 21° E.	−991	+380				
C								991	1580	
	CD	800	243°	S. 63° W.	−363	−713				
D								1354	867	

Length DA $= \sqrt{1354^2 + 867^2} = 1609$ m.

$$\text{Tan } \phi = \frac{1354}{867} = 1{\cdot}5617$$

$$\phi = 57° \ 22'$$

$$\text{W.C.B.} = 270° + 57° \ 22'$$

$$= 327° \ 22'.$$

Example

A traverse ABCD was run round a building site and the following observations made:

Line	*Length (metres)*	*Whole-circle Bearing*
AB	142·8	0° (assumed due north)
BC	127·1	260° 40′
CD	137·3	162° 32′
DA		84° 29′

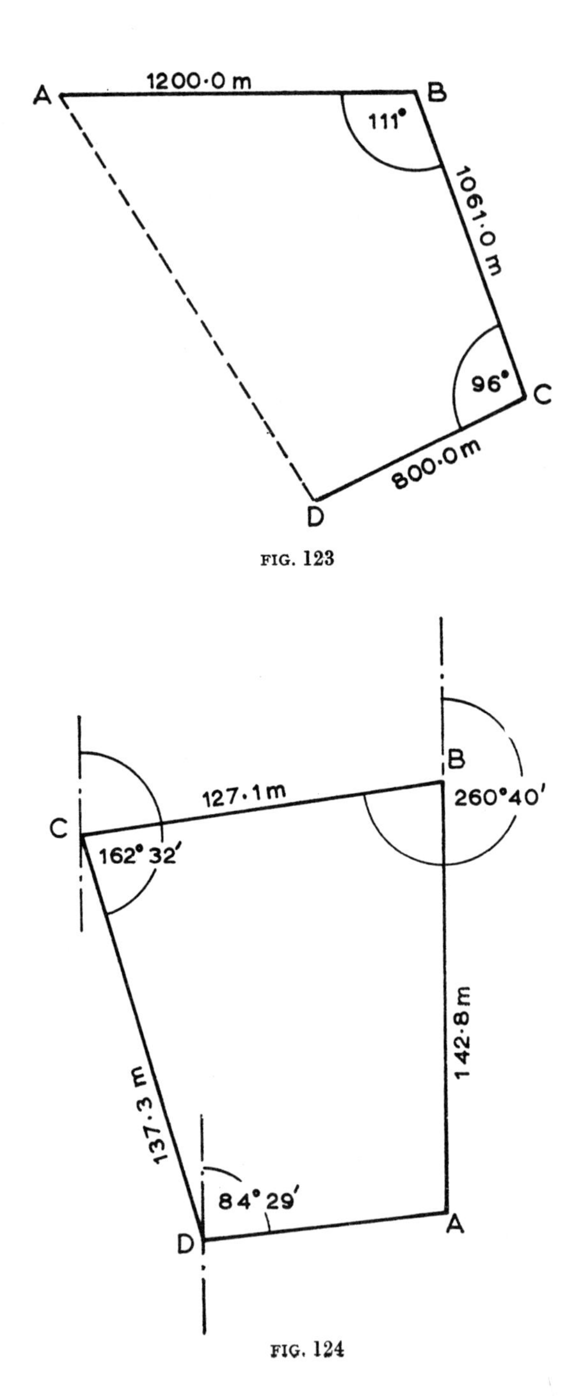

FIG. 123

FIG. 124

Calculate the lengths of AD and BD and set out the traverse in your answer book to a scale of 1 to 400. From your scale drawing measure the lengths of AD, AC and BD. In the actual survey, the latitudes and departures failed to close. Explain briefly, without making calculations, a method of calculating the corrections to be made, and state if the method employed causes any alterations to the bearings of the lines.

Solution (see fig. 124)

To find the lengths of AD and BD, it will first be necessary to find the co-ordinates of B and D.

Sta.	*Line*	*Length (metres)*	*W.C.B.*	*R.B.*	*Lat.*	*Dep.*	*Co-ordinates*			
							N.	*S.*	*W.*	*E.*
A							0	0	0	0
	AB	142·8	0°	0°	+142·8	0				
B							142·8			
	BC	127·1	260° 40′	S. 80° 40′ W.	−20·5	−125·4				
C							122·3		125·4	
	CD	137·3	162° 32′	S. 17° 28′ E.	−131	+41·2				
								8·7	84·2	

$$\text{Length AD} = \sqrt{(8{\cdot}7^2 + 84{\cdot}2^2)} = 84{\cdot}5 \text{ m.}$$
$$\text{Length BD} = \sqrt{[(142{\cdot}8 + 8{\cdot}7)^2 + 84{\cdot}2^2]} = 173{\cdot}3 \text{ m.}$$

Example

In the following traverse table adjust the computed latitudes and departures using Bowditch's Rule, and determine the area in hectares enclosed by the figure ABCDEFA.

Side	*Length (metres)*	*Latitude*	*Departure*
AB	701·40	−602·58	+358·96
BC	249·20	+66·57	+240·14
CD	309·60	+292·96	+100·13
DE	1092·80	+38·88	+1092·11
EF	1278·50	+800·24	−997·08
EA	988·50	−595·09	−789·31

Answer: 958 hectares.

7 CURVE RANGING

THE curves dealt with in this chapter are simple circular curves; transition curves and curves with superelevation, such as are used for railways and major roads, are not dealt with.

Curves may be described in two ways:

(i) by their radius;
(ii) by the angle subtended at the centre by a 100′ chord (American method).

Method (i) will be adopted in this chapter.

PARTS OF THE CURVE

To obtain a smooth change from a straight to a curve, the straight must be tangential to the curve (see fig. 125). The following calculations are based on this requirement.

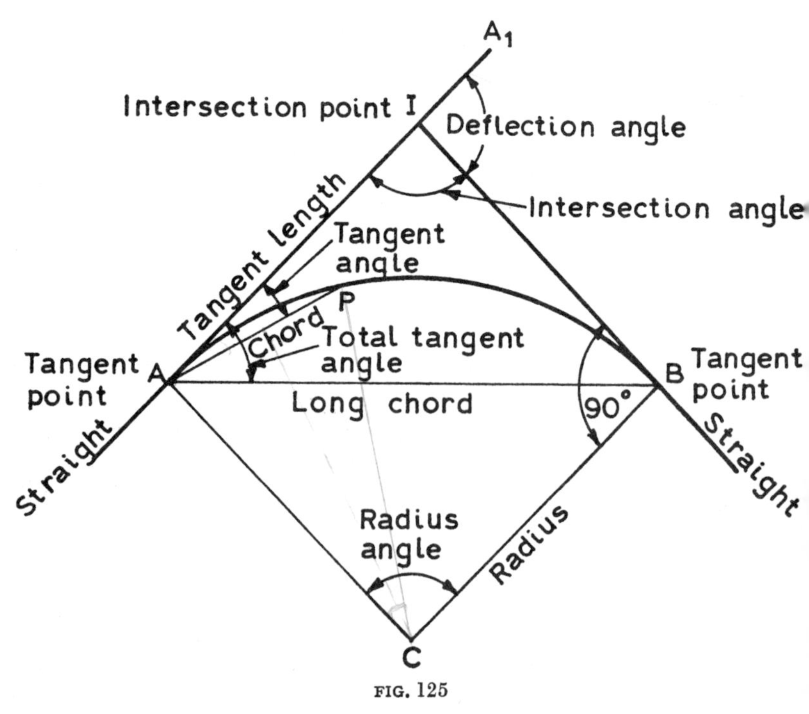

FIG. 125

The names of the parts of the curve are as follows:

Angles

Angle A_1IB is the **Deflection** angle.

Angle AIB is the **Intersection** (or apex) angle.

Angle AIB = 180° – Deflection angle.

Angle ACB is the **Radius** angle = Deflection angle.

Angle IAB is the **Total Tangent angle** = 90° – BAC; and also equals ½ deflection angle.

If AP is any chord, then IAP is its **tangential angle**, and is ½ the angle subtended by the chord (see fig. 125; for proof see p. 125).

Points

Point I is the **intersection point** (abbreviated to I.P.).

Points A and B are the **tangent points** (abbreviated to T.P.).

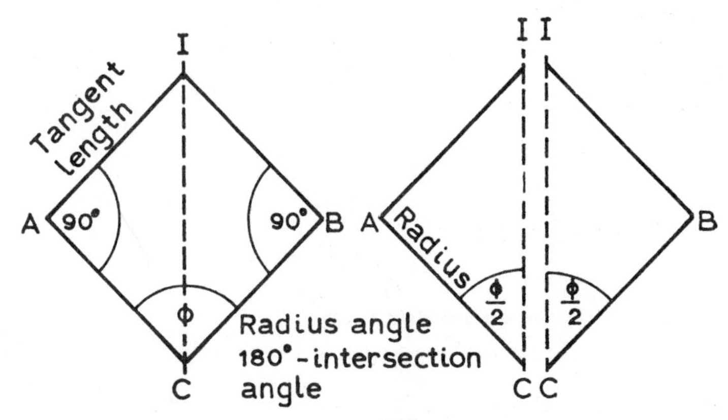

FIG. 126

Lengths

Length along the curve from A to B is the **curve length.**

Straight length from A to B is the **long chord.**

Lengths AI and BI are the **tangent lengths**, and these are always equal. Consider fig. 126, since in the two triangles IAC and IBC, AC and CB are the radii of the curve and equal; IC is common to both triangles and IAC = IBC = 90°; the triangles are identical and AI = BI.

CURVE CALCULATIONS

1. Determine the Deflection and Intersection Angles

These can be determined by measuring from the plan, but it is more accurate to measure the deflection angle by setting out the two straights on the site, fixing the I.P., and measuring the angle AIB with a theodolite. This gives the intersection angle.

The deflection angle = 180° − intersection angle.

2. Decide upon the Length of the Radius

This will depend upon the nature of the site, and the speed and volume of the traffic.

Often on housing and industrial estates the tangent lengths are fixed at some convenient length, the deflection angle measured and the radius calculated from the fixed data.

3. Determine the Tangent Lengths

With the radius given:

$$AI = \text{Radius} \times \tan \tfrac{1}{2} \text{ radius angle (see fig. 126).}$$

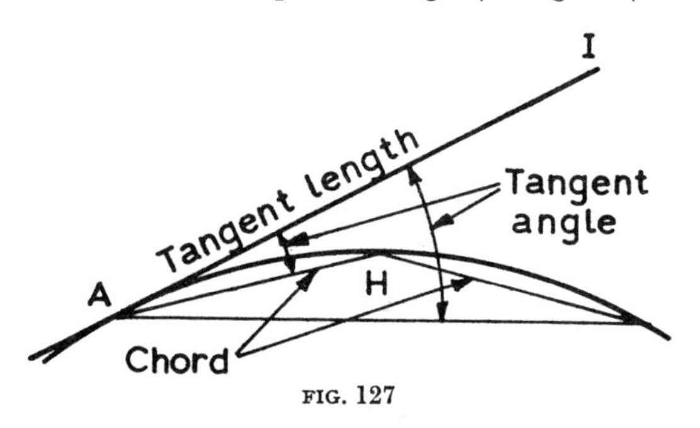

FIG. 127

4. Determine the Length of the Curve

The length of the curve is proportional to the radius angle.

$$\frac{\text{Arc AB}}{\text{Angle ACB}} = \frac{2\pi R}{360°} \text{ where R is the radius.}$$

$$\text{Arc AB} = \text{Length of curve}$$

$$= 2\pi R \times \frac{\text{Angle ACB}}{360°}$$

5. Decide upon the Length of the Chord

The chord length is the **straight** length between two adjacent pegs on the curve (see fig. 127).

For very accurate work the lengths of the chords should not be greater than the $\frac{\text{Radius}}{20}$, but for work where great accuracy is not required (as on estate roads), the length may be made as large as $\frac{\text{Radius}}{10}$. The chord length is assumed to be the same length as the arc, and so must be kept small for accurate work. When inexperienced gangers are laying forms for concrete roads, the chords fixed for setting out form-work should be the same length as the forms, usually about **4** m.

The number of chords required is determined by dividing the chord length into the curve length.

6. **Determine the Tangential Angle** (necessary only when using a theodolite)

The whole tangential angle (total tangent angle) is the angle between the tangent length and the long or main chord. Each sub-chord makes its own tangential angle with the tangent length (see fig. 127). It is

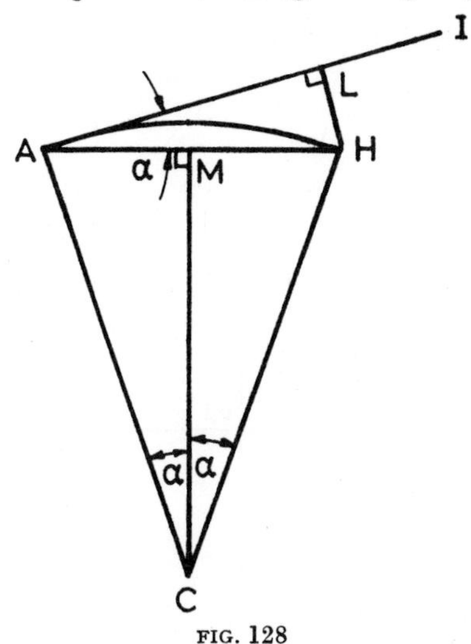

FIG. 128

assumed in the calculations that the length of the sub-chord is the same length as its arc.

In fig. **128**, M is the mid-point of AH. The triangles ALH, AMC are similar since angle AMC = angle ALH = 90°. AC is parallel to LH, hence angle CAM = angle AHL and angle IAH = angle MCA.

Tangential angle for AH (see fig. 128) = IAH = ½ × angle subtended by chord AH.

$$\text{Sin MCA} = \frac{\text{AM}}{\text{AC}} = \frac{1}{2}\,\frac{\text{chord}}{\text{radius}}.$$

$$\text{Therefore}\quad \text{sine of tangential angle} = \frac{1}{2}\,\frac{\text{chord}}{\text{radius}}.$$

Another slightly less accurate method of calculating the tangential angle is as follows:

$$\text{Tangential angle in minutes} = \frac{1719 \times \text{Chord length}}{\text{Radius}}.$$

The formula is derived as follows:

Tangential angle IAH = half angle at the centre of the circle ACH so that angle ACH = 2 × angle IAH. The chord AH, when small compared with the radius, is approximately equal to the arc AH and its length $l = 2R \times 2 \times \frac{IAH}{2R}$.

$$\therefore \quad \text{angle IAH} = \frac{360}{2} \times \frac{l}{2\pi R}$$

$$= \frac{90}{\pi} \times \frac{l}{R} \text{ degrees}$$

$$= \frac{60 \times 90}{\pi} \times \frac{l}{R} \text{ minutes}$$

$$= 1718{\cdot}9 \frac{l}{R} \text{ minutes.}$$

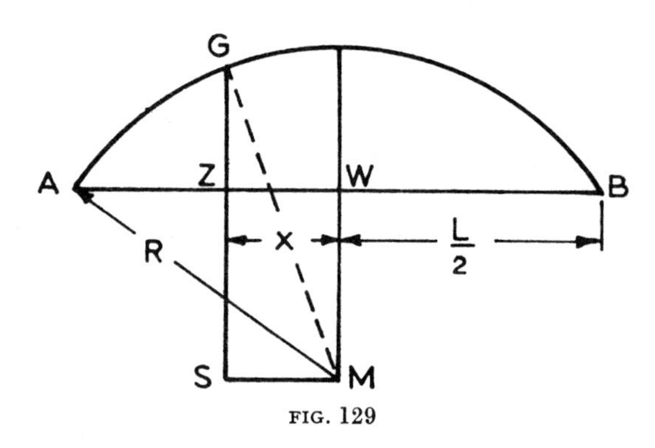

FIG. 129

SETTING-OUT

The setting-out may be done with a chain only or with a theodolite and chain. The chain method should only be used on curves of small radii.

CHAIN METHODS

1. By Right-Angle Offsets from the Main Chord

To find length of offset GZ at distance x from W, mid-point of main chord AB (see fig. 129):

(i) Join MG, let AB = L, then $AW = \frac{L}{2}$.

(ii) In triangle MAW, $MW = \sqrt{R^2 - \left(\frac{L}{2}\right)^2} = K$.

(iii) In triangle GMS, $GS^2 = R^2 - x^2$

$$GS = \sqrt{R^2 - x^2}$$
$$GZ = GS - ZS$$
$$= GS - MW$$
$$= \sqrt{R^2 - x^2} - K.$$

Example

Radius = 500 m; Main chord = 1000 m. Determine the offsets at 100 m intervals from mid-point of main chord.

$$k = \sqrt{500^2 - \left(\frac{1000}{2}\right)^2} = 0$$

when $x = 100$ m Offset $= \sqrt{500^2 - 100^2} = 489{\cdot}9$ m

$x = 200$ m Offset $= \sqrt{500^2 - 200^2} = 458{\cdot}2$ m

$x = 300$ m Offset $= \sqrt{500^2 - 300^2} = 400$ m

$x = 400$ m Offset $= \sqrt{500^2 - 400^2} = 300$ m.

By this method the tangent points are first fixed, and then the mid-point of the main chord is fixed, the x distances set out and the right-angled offsets measured out. It is necessary that the area between the main chord and the curve is clear of obstructions. If the area is not clear, then the following method may be used.

2. **By Perpendicular Offsets from the Tangent Lengths** (see fig. 130)

$$MC^2 = R^2 - D^2$$
$$MC = \sqrt{R^2 - D^2}$$
$$JK \text{ or } x = R - \sqrt{R^2 - D^2}.$$

In this method the tangent lengths are set out, the tangent points fixed, the D distances marked out along the tangent length, and the right-angled offsets measured out. Half the curve is set out from each side.

Example

Radius, 500 m; intersection angle, 90°. Determine the offsets at 100 m intervals from the tangent points.

Tangent length = Radius × tan ½ radius angle
= 500 m × tan ½ 90°
= 500 m × 1
= 500 m.

When $d = 100$ m offset $= 500 - \sqrt{500^2 - 100^2}$
$= 500 - 489{\cdot}9 =$ **10·1 m.**

When $d = 200$ m offset $= 500 - \sqrt{500^2 - 200^2}$
$= 500 - 458{\cdot}2 =$ **41·8 m.**

When $d = 300$ m offset $= 500 - \sqrt{500^2 - 300^2}$
$= 500 - 400 =$ **100 m.**

When $d = 400$ m offset $= 500 - \sqrt{500^2 - 400^2}$
$= 500 - 300 =$ **200 m.**

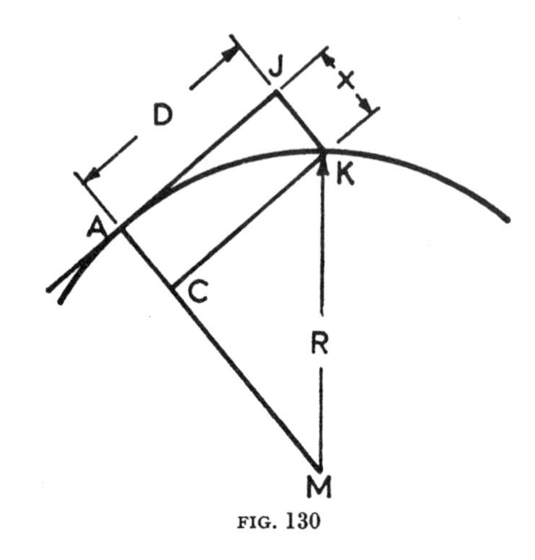

FIG. 130

3. The Method of Extended Chords (see fig. 131)

This method is generally considered to be the most accurate chain method. Procedure is as follows:

Divide the curve into a number of chords of convenient length. Place ranging rods at the tangent points A and B.

Set out along AI a chain to the chord length A*b*. Swing A*b* towards G through the distance *b*G $\left(bG = \dfrac{\text{Chord}^2}{2 \times \text{radius}}\right)$.

Produce AG to F so that GF = AG = chord length.

Swing GF towards H through distance FH $\left(FH = \dfrac{\text{Chord}^2}{\text{Radius}}\right)$.

Produce GH to K and continue as before.

It is rarely possible to divide the curve into a whole number of chords without using fractions of a metre.

It is usual therefore to use, say, 10 m chords and have one chord of a shorter length. The offset for this chord is calculated in a similar manner to the other chords.

Example

Radius, 500 m; intersection angle, 90°. Determine the tangent length, length of curve, number of 50 m chords, length of odd chord and offsets for the 50 m chords and the odd chord.

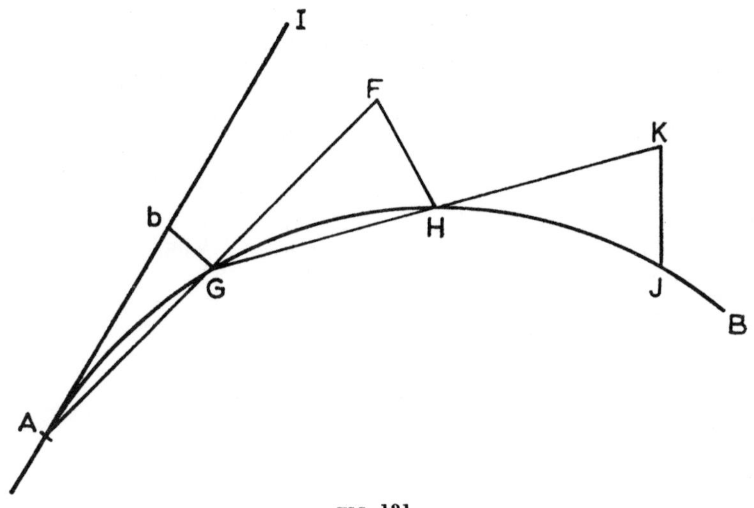

FIG. 131

$$\text{Tangent length} = \text{Radius} \times \tan \tfrac{1}{2} \text{ radius angle}$$
$$= 500 \times \tan \tfrac{1}{2}\, 90^\circ$$
$$= 500 \times 1$$
$$= 500 \text{ m.}$$
$$\text{Curve length} = \frac{2\pi R \times 90^\circ}{360^\circ} = \frac{2\pi \times 500 \times 90}{360^\circ}$$
$$= \mathbf{785{\cdot}4\ m.}$$
$$\text{No. of 50 m chords} = \frac{785{\cdot}4}{50} = \mathbf{15}$$
$$\text{Length of odd chord} = 785{\cdot}4 - 15 \times 50 = 785{\cdot}4 - 750$$
$$= \mathbf{35{\cdot}4\ m.}$$

Offset for odd chord placed at end of curve

$$= \frac{35{\cdot}4^2}{\text{rad.}} = \frac{1253}{500} = \mathbf{2{\cdot}56\ m.}$$

Offset for 50 m chord placed at the beginning of curve

$$= \frac{50^2}{2 \times \text{rad.}} = \frac{2500}{2 \times 500} = \mathbf{2{\cdot}5\ m.}$$

Offset for 50 m chords placed centrally along curve

$$= \frac{50^2}{\text{rad.}} = \frac{2500}{500} = \mathbf{5{\cdot}0\ m.}$$

THEODOLITE METHODS

1. Using One Theodolite

The theodolite is set up over a tangent point, and sighted on to the I.P. with the plates clamped to zero. The instrument is then turned through the tangential angle for the first chord, and a peg inserted on the line of sight at a chord's distance from the tangent point measured with a tape. The instrument is then turned to the tangential angle for the second chord, and a peg inserted in the line of sight at a chord's distance from the previous peg. This procedure is repeated until all the pegs are placed.

The method will be more fully understood by working through the following example.

Example

Carry out the necessary calculations for a 500 m radius curve with a deflection angle of 40°.

Describe the method of setting out the curve.

1. Fix the I.P., measure the deflection angle (given as 40°).
2. Set out along IX and IY the tangent lengths IA and IB.

$$\text{Tangent length} = \text{Rad.} \times \tfrac{1}{2} \tan \text{rad. angle}$$
$$= 500 \times \tan 20°$$
$$= \mathbf{182\ m.}$$

3. Length of curve $= \dfrac{2\pi R \times \text{rad. angle}}{360°}$

$$= \frac{2\pi \times 500 \times 40}{360}$$
$$= \mathbf{349{\cdot}1\ m.}$$

4. Chord length taken as $\dfrac{\text{rad.}}{10} = \dfrac{500}{10} = \mathbf{50\ m.}$

$$\text{No. of chords} = \frac{\text{Length of curve}}{\text{Length of chord}} = \frac{349{\cdot}1}{50}$$
$$= \textbf{6 chords at 50 m and 1 chord at 49·1 m.}$$

5. Tangential angle in minutes for 50 m chords $= \dfrac{1719 \times \text{Chord length}}{\text{Radius}}$

$$= \frac{1719 \times 50}{500}$$
$$= 171{\cdot}9'$$
$$= \mathbf{2° \ 51' \ 54''.}$$

Tangential angle in minutes for 49·1′ chord $= \dfrac{1719 \times 49{\cdot}1}{500}$

$$= 168{\cdot}8'$$
$$= \mathbf{2° \ 48' \ 48''.}$$

As in all surveying and setting out, the fieldwork should be tabulated. A common form is shown below:

Peg No.	*Road Chainage*	*Curve Chainage*	*Tangential Angle*

Suppose the road in the example started at some point S and the length from the start to the T.P. along the road was 1097 m, then the first peg, i.e. at the T.P., would have road chainage, 1097 m, curve chainage 0 m and tangential angle 0°. The second peg would be peg number 2, having road chainage 1147 m, curve chainage 50 m, tangential angle 2° 51′ 54″.

Peg No.	*Road Chainage (metres)*	*Curve Chainage (metres)*	*Tangential Angle*	*Remarks*
1 (T.P.)	1097	0	0°	
2	1147	50	2° 51′ 54°	
3	1197	100	5° 43′ 48″	
4	1247	150	8° 35′ 42″	
5	1297	200	11° 27′ 36″	
6	1347	250	14° 19′ 30″	
7	1397	300	17° 11′ 24″	
8 (T.P.)	1446·1	349·1	20° 0′ 12″	

The tangential angle of peg number 3 is twice times 2° 51′ 54″, which equals 5° 43′ 48″. The tangential angle for peg number 4 is three times 2° 51′ 54″. It will be noted that the tangential angle for the last peg is 20° 0′ 12″, and not 20° as it should be. This will be explained later.

SETTING-OUT THE CURVE

Set up the theodolite at the T.P. at A, sight on to the I.P. at I with the plates clamped to zero. Turn through the tangential angle 2° 51′ 54″ and fix a ranging pole in the line of sight, measure out from the T.P. 50 m, the chord length along the line of sight and insert peg number 2 (see fig. **132**).

Set the plates to read 5° 43′ 48″, and measure 50 m from peg number 2, where the measured 50 m intersects the new line of sight of the theodolite, and insert peg number 3. To fix peg number 4, set off on the plates the angle 8° 35′ 42″ and measure 50 m from peg number 3.

This process is repeated until the last chord, the one of length 49·1 m is set, which should coincide with the T.P. at B.

ADJUSTING THE CURVE

The last peg rarely coincides with the T.P. in practice, this is due to the following reasons:

1. The curve is measured in short **straights** not in short **arcs**.
2. The tangential angle is given to a greater degree of accuracy than

the nearest 20″, i.e. 2° 51′ **54″**, 5° 43′ **48″**—on most theodolites it is only possible to read to the nearest 20″.

3. The formula for tangential angles contains the constant 1719—it is actually 1718·78. This causes each tangential angle to be a little

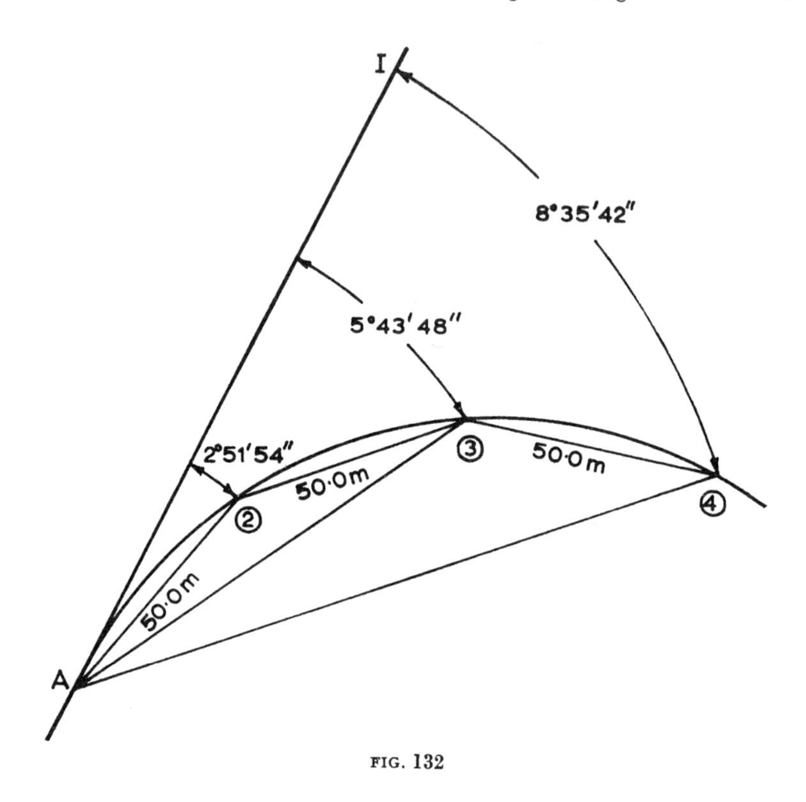

FIG. 132

larger than it should be. In the example, the last tangential angle is 12″ out. It can be seen from fig. 133 that the angle should be 20°, **not** 20° 0′ 12″.

The error in a curve of about $\frac{1}{2}$ mile long may be about 200 mm. The whole adjustment could take place at either end of the curve, but this would tend to destroy the smooth change from straight to curve. It is better to adjust the curve in the centre. Also the length of sight from one end of the curve to the other is great and may lead to further error. If the ground is sloping there will be an accumulative error in measuring slope lengths, which will mount up at the further tangent point B. Half the curve should be set out from **each** tangent point instead of the whole of the curve from one tangent point.

Quite often in practice the curve is set out with the pegs at equal road chainage and, in this case, the tangential angles for the first and last chords are calculated from short chords. In the example the first

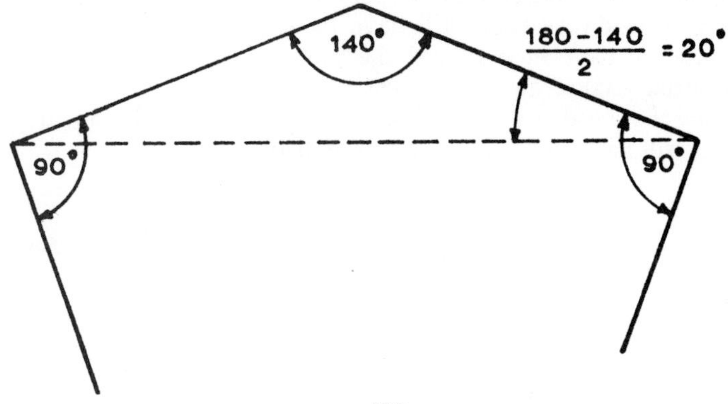

FIG. 133

peg would be required at 1100 m road chainage and the last at 1400 m so that their tangential angles would be

$$\frac{1719 \times 3}{500} = 10{\cdot}3' = 10'\ 20''$$

$$\frac{1719 \times 46{\cdot}1}{500} = 158{\cdot}5' = 2^\circ\ 38'\ 30''$$

The second and following chords are made the normal 50 m length. The curve is set out from T.P.A. using the short (3 m) chord, and then 50 m chords until the pegs are placed just beyond the centre, and the theodolite is moved to T.P.B. and the short chord, 46·1 m, and the 50 m chords are again set out until the centre is passed. There will now be two pegs at the centre and at the ends of the adjoining chords (see fig. 134).

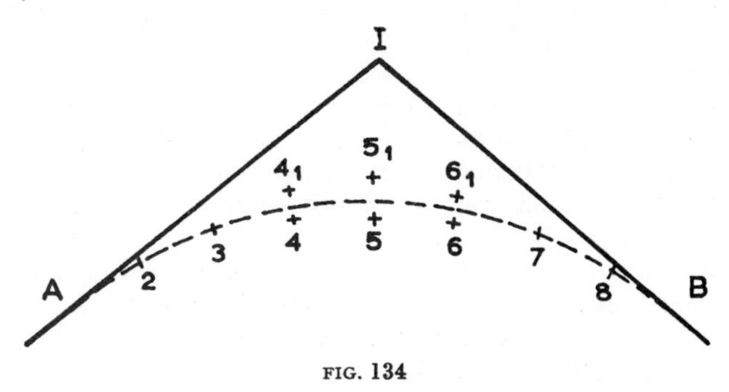

FIG. 134

Suppose the distance apart of the pegs at the centre (5 and 5_1) is 300 mm, insert a peg between the two, i.e. 150 mm from one. Move peg **4 in** towards the radius point half the movement of peg 5, i.e. 75 mm, and move peg **6 out** 75 mm and remove pegs 4, 6, 5 and 5_1.

A new fieldwork table will have to be drawn up for this method because we have now altered the first and last chords' lengths and tangential angles, and also when working from T.P.B. we shall be turning the theodolite in an anti-clockwise direction, and the tangential angles will have to be subtracted from 360°.

Peg No.	*Road Chainage (metres)*	*Curve Chainage (metres)*	*Tangential Angle*	*Remarks*
1 (T.P.)	1097	0	0°	
2	1100	3	0° 10′ 20″	
3	1150	53	3° 02′ 14″	
4	1200	103	5° 54′ 08″	
5	1250	153	8° 46′ 02″	Centre
6	1300	203	11° 37′ 56″	Overlap
9 (T.P.)	1446·1	349·1	360°	
8	1400	303	357° 21′ 30″	
7	1350	253	354° 29′ 36″	
6_1	1300	203	351° 37′ 42″	
5_1	1250	153	348° 45′ 48″	Centre
4_1	1200	103	345° 53′ 54″	Overlap

The angles have been calculated to single seconds; it is not possible to set out to this accuracy with a 20″ instrument so that the calculated values should be rounded off after calculation to maintain as much accuracy as possible.

2. Setting-out Curves using Two Theodolites

Greater speed and accuracy can be obtained by using two theodolites when setting out in this manner.

One theodolite is set up at the T.P.A., set to zero, and sighted on to I.P.I. The other is set up at T.P.B., set to zero, and sighted on to T.P.A.

Both theodolites are turned through the first tangential angle, the chainman moves a ranging rod until it is in the intersection of the lines of sights of the theodolites; a peg is then inserted. The theodolites are then turned through the second tangential angle, the chainman moves a ranging rod until it is in the intersection of the new lines of sight of the theodolites when another peg is inserted (see fig. 135).

This process is repeated until the complete curve is set out. Note that no tape is used for fixing the position of the chord pegs.

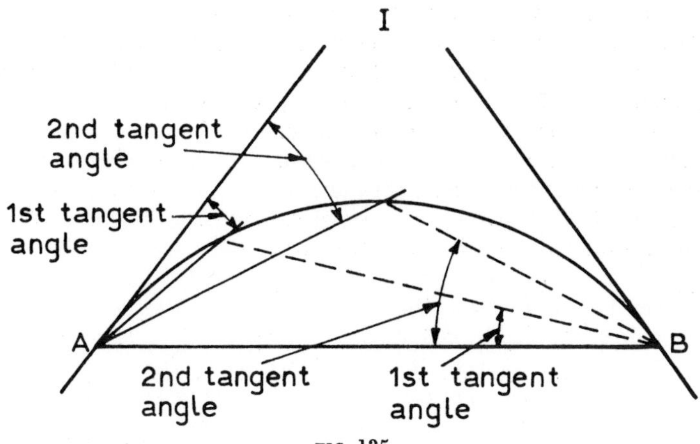

FIG. 135

Obstructions

1. WHEN THE I.P. IS INACCESSIBLE.

The two straights are set out up to the obstruction, and pegs placed at P and Q. The angles XPQ and YQP are read and the distance PQ measured. In fig. 136 let XPQ = 165°, YQP = 155°, and PQ = 110 m. Then IPQ = 180° − 165° = 15° and IQP = 180° − 155° = 25°, and PIQ = 180° − (25° + 15°) = 140°.

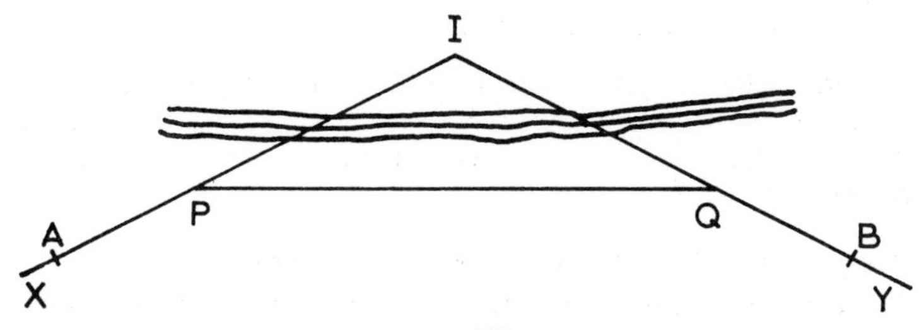

FIG. 136

$$\text{Now} \quad \frac{\text{PI}}{\sin \text{IQP}} = \frac{\text{IQ}}{\sin \text{IPQ}} = \frac{\text{PQ}}{\sin \text{PIQ}}.$$

$$\text{Therefore} \quad \frac{\text{PI}}{\sin 25°} = \frac{\text{IQ}}{\sin 15°} = \frac{110}{\sin 140°}.$$

$$\text{Therefore} \quad \text{PI} = \frac{110 \sin 25°}{\sin 140°} = \frac{110 \times 0{\cdot}4226}{0{\cdot}6428} = \mathbf{72{\cdot}31\ m.}$$

$$\text{IQ} = \frac{110 \times \sin 15°}{\sin 140°} = \frac{110 \times 0{\cdot}2588}{0{\cdot}6428} = \mathbf{44{\cdot}29\ m.}$$

The intersection angle is 140°, and if the radius of the curve is to be 500 m, we know from a previous example that the tangent lengths are 182 m. That is PA = 182 m − 72·31 m = 109·69 m, and QB = 182 m − 44·29 m = 137·71 m. To fix the T.P. at A we must measure 109·69 m from P towards A and we must measure 137·71 m from Q towards Y to fix T.P. at B. The theodolite is set up at A sighted on to P, set to zero and the curve set out in the normal way.

2. WHEN THE LINE OF SIGHT IS OBSTRUCTED BY AN OBJECT.

Set out the curve as far as possible from one T.P., then transfer to the other T.P. and set out the rest of the curve. Alternatively, if it is impossible to set out from B in fig. 137, then as peg 4 cannot be placed from point A, peg 3 is set out from A, then set up the instrument over peg 3, set the horizontal circle to 180° and direct on to point A, transit the telescope and proceed to lay off the remaining tangential angles.

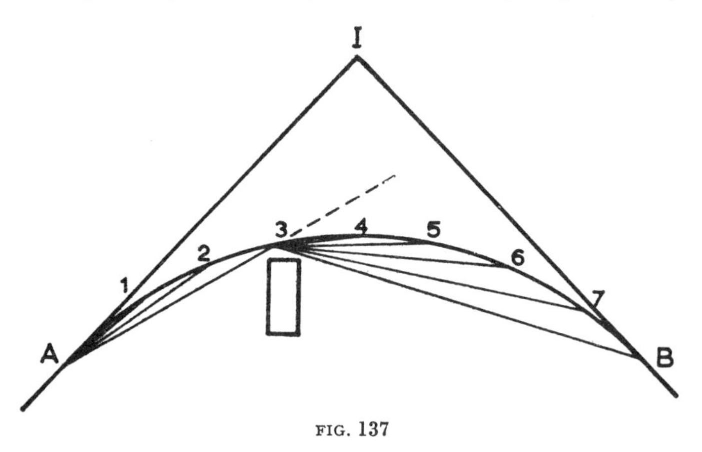

FIG. 137

PRACTICAL NOTES

1. When the I.P. and T.P.s have been located, fix permanent points (see page 163) at the I.P., and at some convenient distance from either side of the T.P.s. During the course of construction of a road or railway the curve centre line pegs may have to be set out half a dozen times, and much time would be wasted on relocating the T.P. and I.P.

2. When the centre line has been set out for the first time, establish permanent points at about every fifth chord on either side of the centre line (see fig. 138).

3. Set out slope stake pegs at every 30 m; the top 150 mm of soil is then removed between the slope stakes and carted away to a soil dump.

4. Reset out the centre line, place level pegs outside the slope stakes so that cuts and fills can be constructed.

5. During cutting and filling operations the centre line may have to be reset out several times to prevent 'drifting' (see fig. 139).

6. If chord lengths of one-tenth of the radius are always used, then the tangential angle will always be 2° 51′ 54″. It will save time to draw up a table for chords' tangential angles for about 50 chords, and paste the table in the cover of the fieldbook.

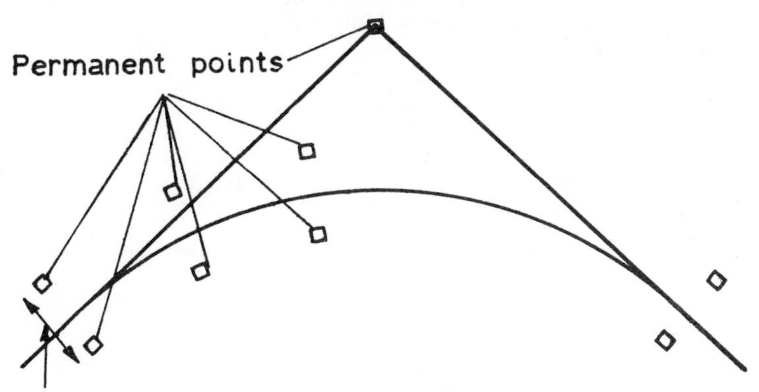

FIG. 138

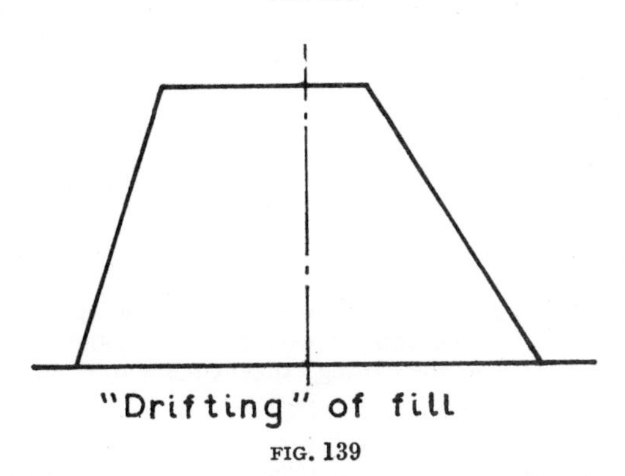

FIG. 139

Example

The lengths and bearings of three successive straights in the final location survey for a railway are:

Line	*Length (metres)*	*Bearing*
PQ	1363·2	122° 39′
QR	839·8	149° 57′
RS	946·2	131° 21′

It is proposed to connect PQ and QR by a curve of 1600 m radius and QR and RS by one of 1000 m radius. If the chainage of the tangent point on PQ is fixed to be 642·2 m find that of the other tangent points.

Solution (see fig. 140)

Deflection angle at Q = 149° 47′ − 122° 39′ = 27° 18′ = 27·3°
Deflection angle at R = 149° 57′ − 131° 21′ = 18° 36′ = 18·6°

$$\text{Length of curve VW} = \frac{2 \times \pi \times 1600 \times 27{\cdot}3^\circ}{360^\circ} = 762{\cdot}4 \text{ m}$$

∴ Chainage of W = 642·3 + 762·4 = 1404·7 m

Now Tangent length QW = 1600 tan 13° 39′ = 388·6 m
∴ Length of WR = 839·8 − 388·6 = 451·2 m
∴ Chainage of R = 1404·7 + 451·2 = 1855·9 m
Now Tangent Length RW = 1000 tan 9° 18′ = 163·8 m
∴ Chainage of X = 1855·9 − 163·8 = 1692·1

$$\text{Length of Curve XY} = \frac{2 \times \pi \times 1000 \times 18{\cdot}6^\circ}{360^\circ} = 324{\cdot}6 \text{ m}$$

∴ Chainage of Y = 1692·1 + 324·6 = 2016·7 m

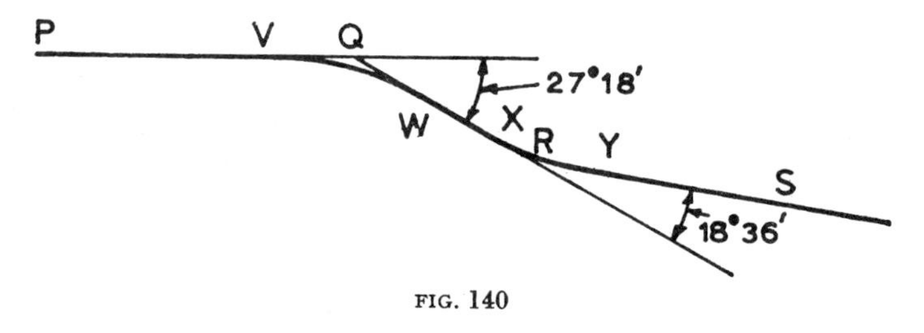

FIG. 140

Example

Define the term 'curve ranging'.

Two straight roads meet at an angle of 127° 30′. If the roads are to be connected by a circular curve of 1500 m radius, find the tangent distance and the length of the curve. One theodolite and a chain are taken to set out the curve in the field using deflection angles. Calculate the deflection angles and show by a sketch on the drawing-paper how the curve would be located. State how the instrument would be manipulated if obstructions to the lines of sight near the closing tangent point were encountered.

Solution

Intersection angle = 127° 30′
∴ Deflection angle = 180° − 127°× 30′ = 52° 30′.
Tangent length = Radius × tan ½ deflection angle
= 1500 × 0·4932
= **739·8 m.**

$$\text{Length of curve} = 2\pi R \times \frac{52^\circ\ 30'}{360^\circ}$$

= **1374 m.**

Example

A circular curve of 500 m radius is to connect two straights which make an internal angle of 156° with one another. Calculate the tangent lengths and the lengths of the long chord and the circular arc. Describe, with sketches, a method of setting out the curve, using a theodolite and a 30 m chain.

Solution

$$\text{Intersection angle} = 156°$$

$$\therefore \quad \text{Deflection angle} = 180° - 156° = 24°.$$

$$\text{Tangent length} = 500 \times \tan 12° = \mathbf{106{\cdot}3\ m.}$$

$$\text{Length of curve} = 2\pi R \times \frac{24°}{360°} = \mathbf{209{\cdot}4\ m.}$$

$$\frac{\frac{1}{2}\text{ length of long chord}}{\text{Radius}} = \text{sine } \tfrac{1}{2} \text{ deflection angle}$$

$$\therefore \quad \text{Length of long chord} = 2 \times \text{Radius} \times \sin \tfrac{1}{2} \text{ deflection angle} = \mathbf{207{\cdot}9\ m.}$$

Example

Indicate with the aid of symbols the steps to be taken to prepare a table of tangential angles for setting out a simple circular curve of radius R metres, between two straight AB and BC, the angle ABC being (180° − I).

Describe briefly an alternative method for setting out the curve.

(Inst. Civ. Engs., Sect. B, October 1949.)

Example

Two straights AI and BI meet at I on the far side of a river. On the near side of the river a point E was selected on the straight AI and a point F on the straight BI and the distance from E to F measured and found to be 34·0 m. The angle AEF was found to be 165° 36′ and the angle BFE 168° 44′. If the radius of a circular curve joining the straights is 200 m, calculate the distance along the straights from E and F to the tangent points.

(Inst. Civ. Engs., Final, Part II, October 1952.)

Answer: EX = 30·22 m
FY = 26·04 m.

Example

At chainage 2946·1 m along the centre line of a proposed road, there is a deflection of 22° 16′ to the *left*. The two straights are to be connected by a circular arc of 500 m radius, which is to be set out by the method of tangential angles.

Calculate:

(*a*) Tangent distance.
(*b*) Tangential angle for 50 m chord.
(*c*) Length of curve: chainage of beginning and end of curve.

Draw up a table of chainages and tangential angles for setting out the curve with pegs at each even 50 m of chainage.

Example

Straight lines AB, BC and CD are to be connected by circular arcs of 300 m and 400 m radius respectively. If the chainage of A is 0, calculate the chainage of D as measured round the curves.

Point	*Bearing*	*Distance (metres)*
A		
	12° 20′	436·8
B		
	25° 40′	512·7
C		
	36° 00′	713·2
D		

8 TACHEOMETRY

TACHEOMETRY is the measurement of the vertical and horizontal distances of a station from the instrument by means of telescopic observations, with the instrument, on to a graduated staff held at the station. Since no chaining or levelling is involved, it forms a rapid method of surveying, and is being increasingly used for preliminary surveys of roads and railways and for contouring. The degree of accuracy will depend upon the range of the telescope, length of sights, etc., but it is at best accurate only to 0·1 m in horizontal distance.

The instrument used is known as a tacheometer, which has been developed from the stadia tube. The stadia tube is rarely used now, but will be described, since it forms a simple introduction to the method.

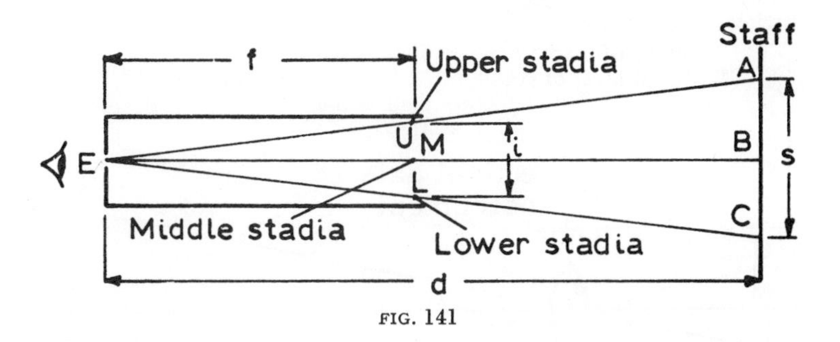

FIG. 141

THE STADIA TUBE

Was invented by James Watt in 1771, and consisted of a tube with a pinhole eyepiece and a diaphragm consisting of three horizontal hairs spaced equidistantly (see fig. 141). The instrument is held horizontally and sighted on to a graduated staff held vertically. In fig. 141 it will be seen that since: triangles EUM and ELM are identical and triangle EUM is similar to triangle EAB and triangle ELM is similar to triangle ECB, EB:AC = EM:UL.

Let EB = d (distance of staff from eye)
AC = s (stadia distance)
EM = f (distance from the eye to the stadia hairs)
UL = i (distance apart of upper and lower stadia hairs)

Then $\frac{d}{s} = \frac{f}{i}$ and $d = s \times \frac{f}{i}$.

$\frac{f}{i}$ is a constant for any particular tube and is known as the **multiplying constant**; the tube is usually made so that $\frac{f}{i} = 100$.

Example

With a stadia tube of multiplying factor 100, the upper stadia reading with 4·500 and the lower 2·500, how far is the staff from the instrument?

Solution

$$d = s \times \frac{f}{i} = 100 \times (4{\cdot}500 - 2{\cdot}500) = 100 \times 2{\cdot}000$$
$$= 200 \text{ m.}$$

Note that s is found to 0·001 m; when this is multiplied by 100, the distance is accurate to 0·1 m.

Example

If the ground level at the instrument is 50·0 m and the instrument is set 2·0 m above ground level, the middle the stadia reading being 3·5, what is the reduced level of the staff station?

Solution

Height of stadia tube = 50·0 + 2·0 = 52·0.
Reduced level of stadia = 52·0 − middle stadia reading
= 52·0 − 3·5
= 48·5 m.

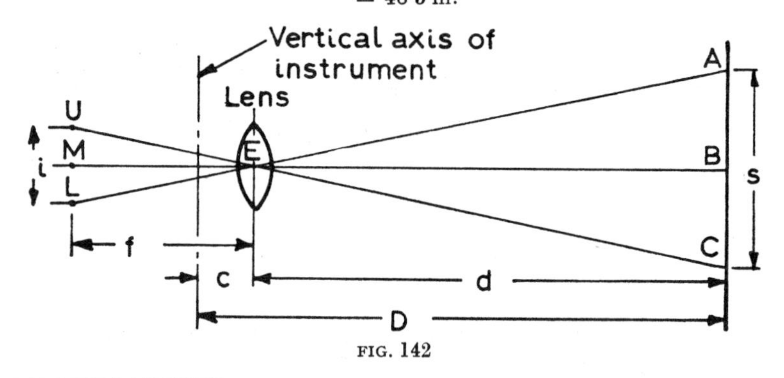

FIG. 142

TACHEOMETER

As Watt's tube was not equipped with lenses, the length of sight was restricted. This disadvantage was overcome by fitting the two stadia wires to the diaphragm of an ordinary transit theodolite (which then becomes a form of tacheometer). The addition of lenses, does not alter the simple geometric principles of the stadia tube (see fig. 142). $\frac{d}{s} = \frac{f}{i}$, but the total distance of the staff from the theodolite, D, $= d + c$. Provided that the object glass is not moved, c is a constant and is known as the **additive constant.** With modern internally focusing telescopes, the additive constant is zero.

Example

The upper and lower stadia readings on a transit theodolite are 4·850 and 2·930 respectively. If the instrument has a multiplying constant of 100, and an additive constant of 1, how far is the station from the instrument? If the height of the instrument is 182·52 m and the axial reading = 3·890, what is the reduced level at the staff station?

Solution

$$\begin{aligned} \text{Distance, D} &= \frac{f}{i} \times s + c \\ &= 100\,(4{\cdot}850 - 2{\cdot}930) + 1{\cdot}0 \\ &= 100 \times 1{\cdot}920 + 1 \\ &= 193{\cdot}0 \text{ m} \\ \text{Reduced level} &= \text{Height of instrument} - \text{axial reading} \\ &= 182{\cdot}52 - 3{\cdot}89 \\ &= 178{\cdot}63 \text{ m.} \end{aligned}$$

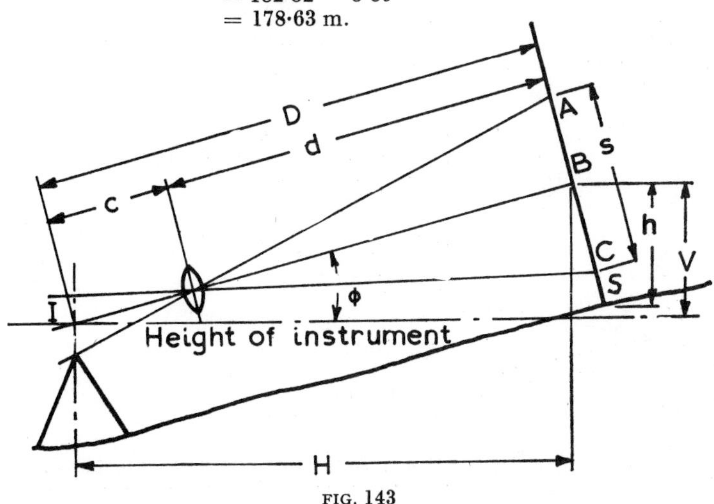

FIG. 143

Inclined Sights (staff held at right angles to the line of collimation)

Horizontal sights are impossible when the instrument and staff stations are at greatly different altitudes and inclined sights have to be taken (see fig. **143**). To enable the staff to be held at right angles to the line of collimation a small telescope may be clamped at right angles to the staff through which the chainman sights on to the theodolite.

$$\text{H} = \text{D}\cos\phi = \left(\frac{f}{i} \times s + c\right)\cos\phi.$$

$$\text{V} = \text{D}\sin\phi = \left(\frac{f}{i} \times s + c\right)\sin\phi.$$

The objections to this method are:

1. It is difficult to hold the staff at right angles to the line of collimation.

2. The distance, H, calculated is the horizontal projection of IB, whereas the distance required is IS.

For these reasons it is not used in practice very much.

Inclined Sights (staff held vertically, see fig. 144)

This is the normal method of holding the staff. Let A′C′ be s and since A′AB is practically a right angle, AC $= s \cos \phi$.

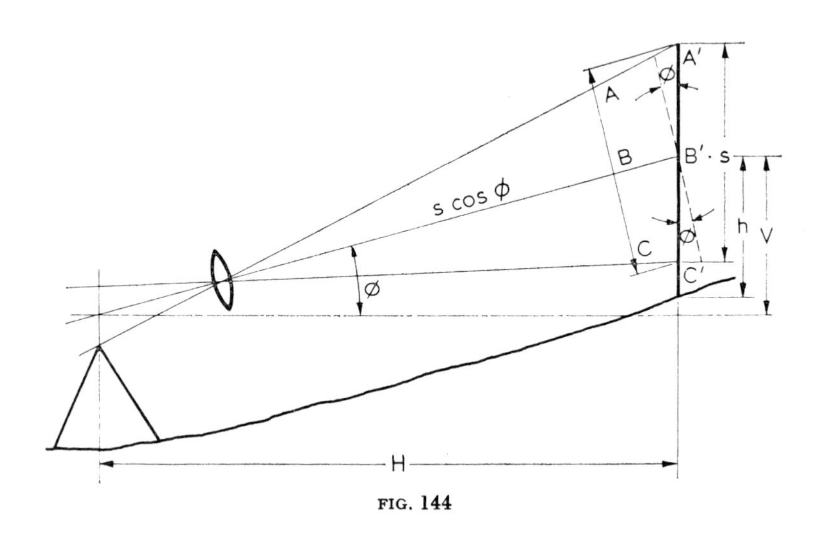

FIG. 144

If we substitute $s \cos \phi$ for s in the previous equation, we shall obtain:

$$H = \left(\frac{f}{i} s \cos \phi + c\right) \cos \phi \qquad V = \left(\frac{f}{i} s \cos \phi + c\right) \sin \phi$$

$$= \frac{f}{i} s \cos^2 \phi + c \cos \phi. \qquad = \frac{f}{i} s \cos \phi \sin \phi + c \sin \phi.$$

Since c is small compared with $\frac{f}{i} \times s$, $c \cos \phi$ may be taken as c and $c \sin \phi$ as 0.

$$\therefore \quad H = \frac{f}{i} s \cos^2 \phi + c. \qquad V = \frac{f}{i} s \cos \phi \sin \phi.$$

Example

A theodolite whose height of instrument level is 182·56 has a multiplying constant of 100 and an additive constant of 1. If the angle of elevation is 10° and the upper, middle and lower stadia readings are respectively 5·0, 3·5 and 2·0, what is the distance of the staff from the station and what is the reduced level at the staff?

$$\begin{aligned} H &= \frac{f}{i} s \cos^2 \phi + c \\ &= 100\,(5{\cdot}0 - 2{\cdot}0) \times 0{\cdot}9848^2 + 1 \\ &= 300 \times 0{\cdot}97 + 1 \\ &= \mathbf{292\ m.} \end{aligned}$$

$$\begin{aligned} v &= \frac{f}{i} s \cos \phi \sin \phi \\ &= 100\,(5{\cdot}0 - 2{\cdot}0) \times 0{\cdot}9848 \times 0{\cdot}1736 \\ &= 300 \times 0{\cdot}9848 \times 0{\cdot}1736 \\ &= 51{\cdot}28\ \text{m.} \end{aligned}$$

Reduced level of ground at staff

$$\begin{aligned} &= \text{Ht. of inst.} + V - h \\ &= 182{\cdot}56 + 51{\cdot}28 - 3{\cdot}5 \\ &= \mathbf{230{\cdot}34\ m.} \end{aligned}$$

Example

A theodolite with a multiplying constant of 100 and an additive constant of zero sights on to a B.M. 75·0 with an angle of depression of 10°. Determine the height of the instrument when the stadia readings are 8·0, 5·0 and 2·0.

$$\begin{aligned} v &= \frac{f}{i} s \cos \phi \sin \phi \\ &= 100\,(8{\cdot}0 - 2{\cdot}0) \times 0{\cdot}9848 \times 0{\cdot}1736 \\ &= 102{\cdot}56\ \text{m.} \\ \text{Ht of inst.} &= \text{B.M.} + v + h \\ &= 75{\cdot}0 + 102{\cdot}56 + 5{\cdot}0 \\ &= 182{\cdot}56\ \text{m.} \end{aligned}$$

If the same instrument now sights on to a station with an angle of depression of 15° and the stadia readings are 8·5, 5·7 and 2·9, find the distance of the staff station from the instrument station and its reduced level.

$$\begin{aligned} H &= \frac{f}{i} s \cos^2 \phi + 0{\cdot}0 \\ &= 100\,(8{\cdot}5 - 2{\cdot}9) \times 0{\cdot}9659^2 + 0{\cdot}0 \\ &= 522{\cdot}5\ \text{m.} \end{aligned}$$

$$\begin{aligned} v &= \frac{f}{i} s \cos \phi \sin \phi \\ &= 100\,(8{\cdot}5 - 2{\cdot}9) \times 0{\cdot}9659 \times 0{\cdot}2588 \\ &= 140\ \text{m.} \end{aligned}$$

Reduced level of staff station

$$\begin{aligned} &= \text{Ht of inst.} - V - h \\ &= 182{\cdot}56 - 140{\cdot}0 - 5{\cdot}7 \\ &= 36{\cdot}86\ \text{m.} \end{aligned}$$

METHOD OF SURVEY

1. **Fix stations**—this depends upon the nature of the survey. A route survey requires station pegs placed at points of maximum visibility. If, however, the purpose of the survey is to contour rapidly an area within an existing traverse survey, the station pegs will have already been established. The survey of a large area such as a valley reservoir site is better made by laying out a framework of connected triangles or polygons by normal traversing, and taking the contours in each figure afterwards.

2. **Set up the instrument over the first station,** peg A (see fig. 145), and sight on to a staff held on a B.M. or peg of known reduced level. The height of the instrument can be determined, and by measuring down from the trunnion axis of the instrument to the top of the peg and to the ground, the reduced level of the peg and ground can be determined.

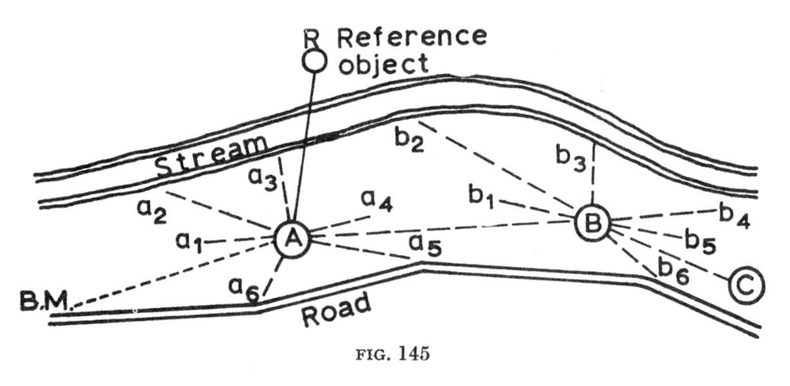

FIG. 145

3. **Set instrument,** clamped to zero, with the telescope lying in the magnetic meridian or sighted on to some reference object, R. The object is chosen so that the bearing of AR can be determined later.

4. **From A 'sideshots' are taken** on to level staffs held at a_1, a_2, a_3, etc. In each case the angles RAa_1, RAa_2, etc., are noted, the vertical angles and stadia readings also being noted. The staff stations should be chosen so that they are at points of importance such as boundaries of fields, stream banks, corners of buildings, points of change of ground slope, etc. It is advisable for speed to have three or four staffmen directed by the surveyor, who may leave the instrument reading to a trained assistant. The staffmen should be told at each new set-up of instrument where they are to hold the staffs, otherwise time will be wasted by the surveyor continually directing the staffmen. The number of sideshots will depend upon the nature of the ground, but generally 20 to 30 are required at each station.

TABLE

TACHEOMETRIC TRAVERSE FIELD-BOOK

Inst. Sta.	Staff Sta.	Height of Trunnion Axis above Sta.	Angle	Bearing	Vert. Angle	Stadia Readings	Stadia Difference	Axial Reading	Horiz. Dist. H.	Vert. Dist. V.	Rise	Fall	Reduced Level		Remarks
													Inst. Axis	Staff Staff Sta.	
A	—	1·600	RA—	210°	−10°	5·000 1·000	4·000	3·000		−68·40		71·40	146·40	75·00	B.M.
	a_1	,,	R A a_1	10°	−15°	4·513 0·976	3·537	2·744	330·0	−88·42		91·16		55·24	Stream bank
	a_2	,,	R A a_2	70°	0°	4·876 1·319	3·557	3·098	355·7	0·0		3·10		145·30	Fence corner
	a_3	,,	R A a_3	170°	+10°	4·502 0·613	3·889	2·558	377·2	+66·51	63·95			210·35	Corner of building
	a_4	,,	R A a_4	185°	+5°	4·498 2·846	1·652								
	B	,,	RAB	60°	+10°	4·076 1·031	3·045	2·554	295·3	+52·07	49·52			195·92	
B	A	1·040			−9° 36′	4·563 1·578	2·985	3·070	290·2	−49·09		52·16	196·96	144·80	
	b_1	,,	A B b_1	10°	−5° 6′	4·892 1·231	3·661	3·062	363·2	−32·42		35·48		161·48	
	b_2	,,	A B b_2	45°	+5°	4·845 0·631	4·214								

5. When all the sideshots are taken from A, the instrument is sighted on to a staff held at station B when the angle RAB, vertical angle and stadia readings are noted. The instrument is then moved to and set up over B, clamped to zero and sighted back to A. Sideshots to b_1, b_2, b_3, etc., are taken, noting the horizontal angles ABb_1, ABb_2, etc., and the vertical angles and stadia readings. Instead of clamping to zero when sighting back on to A, the surveyor may clamp the plates to 360° – angle RAB, so that when the instrument is then turned to 360° (i.e. zero) it is lying in the same meridian as AR, or if AR is the magnetic meridian the instrument is in the magnetic meridian.

The surveyor should take with him on the survey a map of the area mounted on plyboard on which he can make notes of the instrument and staff stations, and sketch in minor details not fixed by staff stations.

FIELD-BOOK

There are various types of field-books used in tacheometry, and one type is given on p. 147, the first nine columns being filled in in the field and the remaining columns filled in later from calculations made in the office. The calculations are similar to those described in previous examples, and should be carried out again by the student; some of the calculations have not been carried out, and the student is advised to calculate and insert the results. The multiplying constant is 100 and the additive constant, zero.

The reduced level of B.M. (75·00) is obtained from an Ordnance Survey map or from a previous survey. The height of the B.M. station below the instrument axis is calculated to be 68·40 + 3·00, i.e. 71·40, and this figure is entered into the Fall column. The height of the instrument axis is therefore 75·0 + 71·40 = 146·40. From this figure the height of a_1 is calculated, i.e. 146·40 – fall of 91·16 = 55·24. The a_3 staff station is **above** the instrument, hence its reduced level is 146·40 **plus** 63·95 = 210·35.

In a like manner the reduced level of station B is determined as 195·92. Since the height of the trunnion axis is 1·04 above the station, the height of the instrument is 195·92 + 1·04 = 196·96. As a check on the work, the reduced level of A is determined from B and found to be 144·80. Now, the height of the instrument when at A was 146·40 and was 1·60 above the station, the reduced level of the station was therefore 146·40 – 1·60 = 144·80, the figure previously obtained, and the levels check.

When the book is completed (the lengths of the lines and bearings being known), a traverse book is drawn up as described in Chapter 6. The traverse is then plotted and the staff stations fixed, together with a note of their reduced levels; contour lines are plotted by interpolation, as described on page 47.

ADDITIONS TO THE THEODOLITE AND OTHER TYPES OF TACHEOMETERS

In an endeavour to increase the accuracy and reduce the labour in calculations in tacheometry, various modifications have been made to the instrument, some of which are described below.

THE ANALLATIC LENS (PORRO LENS)

An additional lens, known as the anallatic lens, is sometimes fitted to theodolites to eliminate the additive constant and thus simplify the calculation slightly (see fig. **146**). It has the disadvantage that it cuts down the amount of light entering the eyepiece, and therefore the length of sight of the instrument, and adds to the cost of the theodolite.

Rays UP and LR parallel to MO are refracted by the anallatic lens to its principal focus F, and so on to object glass O at *n* and *m*; they are bent here and pass on to C and A.

The focusing of the telescope is achieved by moving the diaphragm and the eyepiece, the lenses only remaining stationary. Thus the direction of rays UP and LR are parallel, and will always pass through F and strike the object glass at *n* and *m* and emerge in the direction *n*C and *m*A.

The telescope is so adjusted that *m*A and *n*C if produced backwards would meet at D on the vertical axis. If angles ADB and BDC are represented by ϕ, then:

$$\frac{AB}{BD} = \tan\phi \text{ and } \frac{BC}{BD} = \tan\phi$$

$$\frac{AB}{BD} + \frac{BC}{BD} = 2\tan\phi \text{ and } \frac{AB + BC}{BD} = 2\tan\phi$$

$$\frac{AC}{BD} = 2\tan\phi \text{ and } \frac{BD}{AC} = \frac{1}{2\tan\phi} = \tfrac{1}{2}\cot\phi.$$

So that if BD $=$ D and AC $= s$ (staff intercept) then $\frac{D}{s} = \frac{1}{2}\cot\phi$, therefore D $= s \times \frac{1}{2}\cot\phi$.

Now, as angles BDA and BDC never change, $\frac{1}{2}\cot\phi$ is constant and may be any desired value by adjusting the stadia lines so that $\frac{1}{2}\cot\phi = 100$ and the distance from the staff to the instrument $= 100 \times s$ and no additive constant is required.

Stadia lines are difficult to adjust and in an instrument where $\frac{f}{i} = 100$, any error in i will cause a considerable error in readings. When the anallatic lens is inserted the multiplier is then $\frac{1}{2}\cot = (=100)$, and this is affected by the position of the lenses as well as the stadia lines. Any error in the stadia lines may be adjusted by moving the anallatic lens, which is provided with a screw for this purpose.

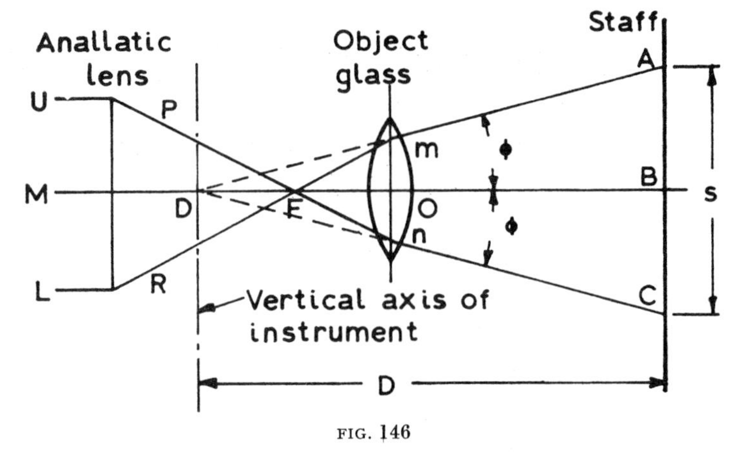

FIG. 146

To make this adjustment, place pegs at exactly 200 m apart and at about the same level. Set up the instrument over one of the pegs, clamp the vertical circle to zero, sight the telescope on to a staff held on the other peg. For an instrument with a factor of 100, the stadia intercept should be 5·0. If it is not, move the anallatic lens until it is.

BEAMAN STADIA ARC

This is a specially graduated scale fixed to the vernier of the vertical circle (see fig. 147). The scales give the stadia coefficient for the vertical angle used. The V scale giving values of $\sin \phi \cos \phi$, the zero reading being given an arbitrary value of 50 to avoid the use of negative quantities. The H scale giving values of $100(1 - \cos^2 \phi)$.

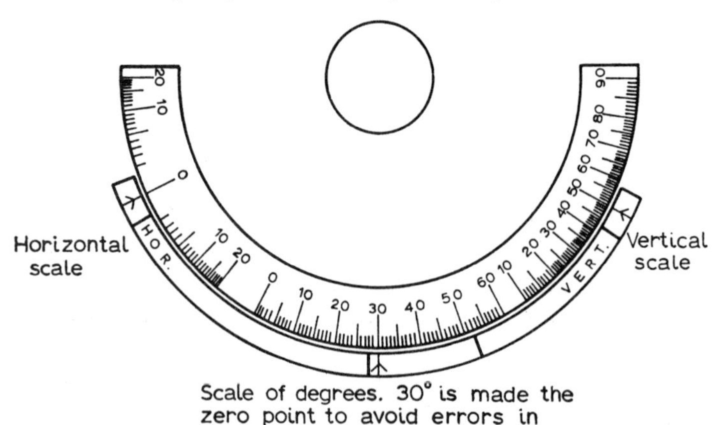

FIG. 147

Example

If the stadia readings are 4·800, 2·500 and 0·200 respectively on an instrument fitted with an anallatic lens and multiplying constant of 100, find the vertical distance of a staff station above the instrument when the V scale reading is 58·8. If the corresponding reading on the H scale is 0·8, what is the horizontal distance to the staff and the height of the staff station above the instrument?

$$\begin{aligned} V &= \frac{f}{i} s \cos\phi \sin\phi \\ &= s \times 100 \cos\phi \sin\phi \\ &= (4{\cdot}800 - 0{\cdot}200) \times (58{\cdot}8 - \text{zero reading, } 50) \\ &= 4{\cdot}600 \times 8{\cdot}8 \\ &= 40{\cdot}48 \text{ m.} \\ H &= 100s \cos^2\phi \\ &= 100s - 100s\,(1 - \cos^2\phi) \\ &= 100 \times 4{\cdot}600 - 4{\cdot}6 \times 0{\cdot}8 \\ &= 460 - 3{\cdot}68 \\ &= 456{\cdot}32 \text{ m.} \end{aligned}$$

DIRECT-TEACHING TACHEOMETER

A direct-reading tacheometer has stadia hairs controlled by cams which cause them to move as the angle of elevation or depression alters.

Thus staff intercept of one stadia line to axial line × 100 = Distance, and the staff intercept of the other stadia line to axial line × 100 = Height.

Example

A theodolite with eyepiece focusing is used for stadia measurements. Obtain an expression for the distance 'D' of a vertical staff from the trunnion axis if the line of sight of the telescope is horizontal.

The observations below were made to determine the two constants for the above instrument:

Distance 'D'	*Reading on Staff*	
	Lower Wire	*Upper Wire*
100 m	3·620	4·610
200 m	2·980	4·970

Using the expression mentioned above, find the value of the constants.

Solution

$$\text{Distance D} = \frac{f}{i} \times s + c$$

$$200 = \frac{f}{i} \times (4{\cdot}970 - 2{\cdot}980) + c$$

$$200 = \frac{f}{i} \times 1{\cdot}99 + c \quad . \quad . \quad . \quad (1)$$

$$100 = \frac{f}{i} \times (4{\cdot}610 - 3{\cdot}620) + c$$

$$= \frac{f}{i} \times 0{\cdot}99 + c \quad . \quad . \quad . \quad (2)$$

(1)–(2)

$$100 = \frac{f}{i} \times 1{\cdot}0$$

$$\therefore \quad \frac{f}{i} = 100.$$

Substitute in (2) $100 = 100 \times 0{\cdot}99 + c$

$$\therefore \quad c = 100 - 99$$

$$c = 1$$

Example

What is meant by tacheometry? State which of the following fractions represents the probable error of horizontal distances when measured by tacheometrical methods: $\frac{1}{120}$, $\frac{1}{600}$, $\frac{1}{10\,000}$; and give reasons for your answer.

A tacheometer is set up at A (reduced level 399·5 m, height of instrument 1·72 m) and readings are taken to two stations B and C as follows:

Staff Station	*Bottom Wire*	*Middle Wire*	*Top Wire*	*Vertical Angle*
B	1·12	2·89	4·66	+3° 47′
C	0·37	2·62	4·87	−5° 26′

Calculate the height above sea-level of B and C, and their horizontal distances from A. Take the multiplying constant as 100 and assume the instrument anallatic. The staff is held vertically in each case.

Solution

(i) Probable error: The maximum length of sight with a tacheometer is about 250 m, with an instrument of multiplying constant 100 the staff intercept will be about 2·5 m. The staff can be read only to the nearest millimetre which when multiplied by the constant means a distance of 0·1 m from the instrument. At a distance of 250 m it is difficult to tell whether the sight is 0·001 ± 0·0002. It is therefore impossible to say whether the distance, for example, should be 249·8 m or 250·2 m. The error is likely to be 0·2 m in 250 m, or $\frac{1}{1250}$. The nearest to this in the fractions given in the question is $\frac{1}{600}$.

(ii) Horizontal distances of B and C:

$$\text{Horiz. dist. of B from A} = \frac{f}{i} s \cos^2 \phi$$
$$= 100\ (4{\cdot}66 - 1{\cdot}12) \cos^2 3^\circ\ 47'$$
$$= 352{\cdot}5 \text{ m}$$
$$\text{Horiz. dist. of C from A} = 100\ (4{\cdot}87 - 0{\cdot}37) \cos^2 5^\circ\ 26'$$
$$= 446{\cdot}2 \text{ m.}$$

(iii) Reduced level of B and C:

$$\text{Reduced level of B} = \text{Ht of inst.} + v - h$$
$$= (399{\cdot}5 + 1{\cdot}72) + \frac{f}{i} s \cos \phi \sin \phi - 2{\cdot}89$$
$$= 401{\cdot}22 + 100 \times (4{\cdot}66 - 1{\cdot}12) \cos 3^\circ\ 47' \sin 3^\circ\ 47' - 2{\cdot}89$$
$$= 401{\cdot}22 + 23{\cdot}30 - 2{\cdot}89$$
$$= 421{\cdot}63.$$
$$\text{Reduced level of C} = 401{\cdot}22 - 100\ (4{\cdot}87 - 0{\cdot}37) \cos 5^\circ\ 26' \sin 5^\circ\ 26' - 2{\cdot}62$$
$$= 401{\cdot}22 - 42{\cdot}42 - 2{\cdot}62$$
$$= 356{\cdot}18 \text{ m.}$$

Example

Obtain an expression for the horizontal distance D of a vertical staff from a theodolite provided with stadia lines, if the line of sight is inclined at ϕ to the horizontal. Explain also how the constants of the expression could be obtained.

From the following data find the horizontal distance and the reduced level of a point B when the staff is held vertically at B, the height of the instrument above a point A is 1·862 m and the reduced level at A is 101·32 m and the angle ϕ is 12°: the staff readings in this case being 4·226 m, 2·619 m and 1·012 m respectively.

Assume that the constants are 100 m and +1·3 m.

Solution

$$\text{Horiz. dist. of B from A} = \frac{f}{i} s \times \cos^2 \phi + c$$
$$= 100\ (4{\cdot}226 - 1{\cdot}012) \cos^2 12^\circ + 1{\cdot}3$$
$$= 307{\cdot}7 + 1{\cdot}3 \text{ m}$$
$$= 307{\cdot}0 \text{ m}$$
$$\text{Reduced level of B} = (101{\cdot}32 + 1{\cdot}86) + [100(4{\cdot}226 - 1{\cdot}012) \cos 12^\circ \sin 12^\circ] - 2{\cdot}62$$
$$= 103{\cdot}18 + 65{\cdot}36 - 2{\cdot}62 \text{ m}$$
$$= 165{\cdot}92 \text{ m.}$$

Example

A level fitted with stadia lines has a multiplying constant of 100 and a zero additive constant. This has been used to obtain the readings (given in the table) of the cross-section of a stream. Calculate the flow of the water in m^3 per sec. if the average velocity is $2\frac{1}{2}$ m per sec.

Note.—The instrument was aligned on the section line.

Position of Staff	*Stadia Readings*			*Remarks*
	Lower m	*Middle m*	*Upper m*	
1	1·48	1·60	1·72	Near bank water level.
2	3·60	3·80	4·00	
3	4·34	4·70	5·06	
4	3·47	4·00	4·53	
5	0·96	1·60	2·24	Far bank water level.

9 PLANE-TABLE SURVEYING

THE use of a plane table enables a survey to be rapidly made. Although the result is not so accurate as that obtained by theodolite or chain survey, the method is sometimes used for filling in the details of a theodolite traverse.

The plane table (see fig. 148) consists of a drawing-board fitted on to a tripod. When unclamped by the wing-nut on the tripod, it rotates horizontally through 360°. The tripod may with advantage be provided with levelling screws. A sheet of strong drawing-paper is attached to the drawing-board, and on it is placed the following:

(i) **A spirit bubble** to enable the board to be levelled.

(ii) **A trough compass** to orientate the survey.

(iii) **An alidade.** This in its simplest form is a straight-edge with a scale marked on and sighting vanes fixed to its ends. A more elaborate alidade consists of a telescope which can revolve through 30° in elevation or depression, and is fitted with a Beaman stadia arc.

METHOD OF SURVEY

1. **Fix Stations**—If the purpose of the survey is to fill in details of an existing theodolite traverse, then the stations will have been fixed. If no stations are fixed, choose first two stations which are intervisible and command a wide view of the survey area. Other stations are fixed so that they locate important points, such as changes of direction of boundaries, gates, etc., and are visible from the first two stations (A and B in fig. 149).

2. **Measure the Line AB**—This line is known as the base line of the survey, and care should be exercised in measuring it, since if there is an error it will affect the whole of the survey.

3. **Set Up the Table**—The table is set up over station A, levelled up by the foot screws, and then rotated until the trough compass, which is placed parallel to one edge, has its needle over the centre of its scale. A needle or pin is stuck in the paper vertically over the station peg A. The alidade is placed against the needle and rotated until it sights on B. A line is drawn along the edge of the alidade and the length AB is plotted to scale from the needle, and the two ends of the line are marked *a* and *b*.

4. **Fixing the Other Stations on the Plan**—The alidade, still placed against the needle, is turned and sighted on to C. A line is drawn from *a* towards C along the alidade edge. Lines are similarly drawn

from *a* to the other stations. When all the stations have been 'lined', the table is then moved to and set up over station B so that *b* is vertically over station B. A needle is then placed through *b*, and the

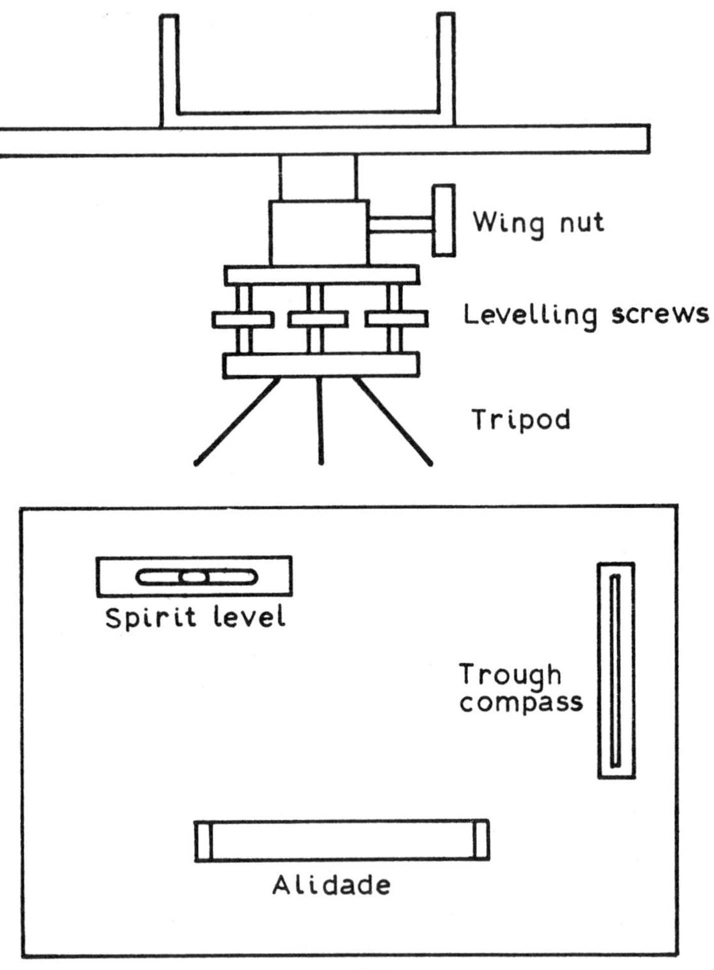

FIG. 148

alidade is placed against it and on the line *ab*. The table is rotated until station A is in the sights of the alidade and then clamped. The alidade is then sighted on to station C and a line drawn towards it from *b*; where this line intersects the line drawn towards C from *a*, the intersection is the relative position of C on the survey. The alidade is turned on to the other stations, lines drawn, and the intersections of these lines with the corresponding lines drawn from *a* fixes the position of the stations.

Instead of drawing lines from *a* and *b* to the stations, short lines about 30 mm long, known as *repère* marks may be drawn at the edge of the paper, in line with the stations. This will help to keep the paper

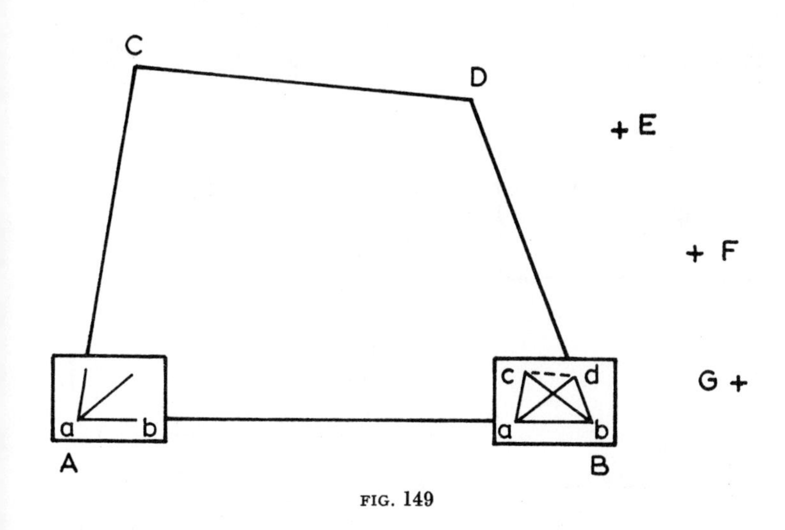

FIG. 149

from becoming a confused mass of lines. The *repère* marks should have a note stating to which stations they were sighted.

5. To Fix Stations not Visible from both Stations A and B— Suppose stations E, F, G, etc., cannot be seen from A, but can be seen from both B and D. *Repère* marks are drawn from C towards these stati ons, the table is then set up at B, orientated, and then sighted on to the stations.

RESECTION, THREE-POINT PROBLEM

New stations are usually plotted from the previous station, but it is sometimes necessary to set up the table in a position which has not been previously plotted. This point may be located on the survey in the following manner:

Drive in a peg to mark the station X.

Set up the table over the station, and place a sheet of tracing-paper on it.

Fix the point *x* on the paper by means of a plumbing fork (see fig. 151).

Choose three points already located upon the survey, and draw rays from x through these points. Now place the tracing-paper over the survey so that the three rays pass through the corresponding points. Point x may then be pricked on to the survey.

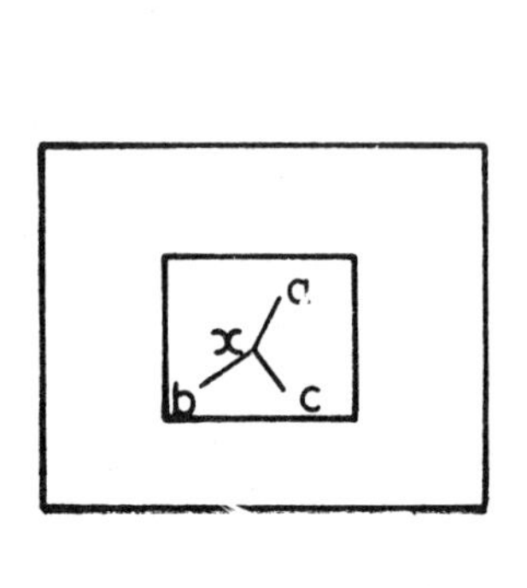

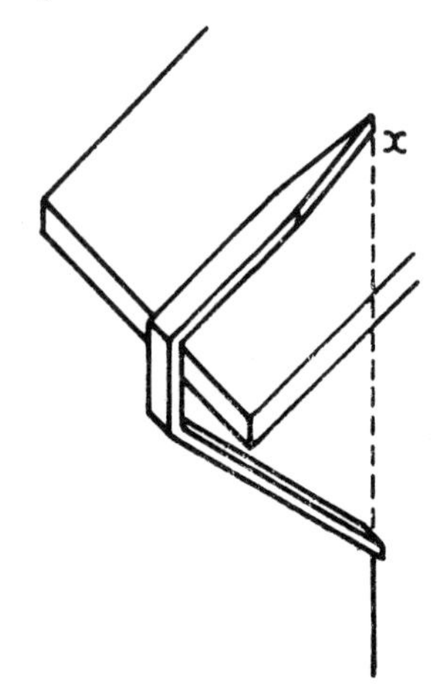

FIG. 150

ADVANTAGES

1. It is a very rapid method of surveying.

2. The map is drawn by the surveyor himself, not by a draughtsman from his notes—a method which sometimes, through misinterpretation of the notes, leads to errors in plotting.

3. The surveyor whilst plotting can see at a glance if he has overlooked some essential part of the plan and can also sketch in minor details.

DISADVANTAGES

1. The method is not very accurate.

2. It can only be used in countries with a reliable dry climate.

ERRORS

The accuracy of the work is mainly governed by the scale of the plan, the larger the scale the smaller the error. The sources of error are as follows:

(i) Shrinkage and warping of the paper. This is the main cause of error.

(ii) The table not being level when taking sights.

(iii) Inaccurate setting of table so that station marked on plan is not over station on the ground. This may be overcome by using a plumbing fork (see fig. 150).

10 SETTING OUT AND MEASURING BUILDINGS

SETTING OUT BUILDINGS

To locate a building either two base lines or one base line and a fixed point are necessary. In setting out houses, the base line may be the building line, as defined by the local authority, or the road line. On an industrial estate the base line may be the grid lines (see page 2). It is also necessary to have a site plan with the base line drawn on it together with the ground plan of the proposed buildings. The measurements to be set out are then scaled off the plan. It is generally quicker and more accurate to do all setting out with a theodolite and steel tapes, and when setting out steel or reinforced-concrete framed buildings it is essential to use this method. However, for buildings where less accuracy is required linen tapes only are generally used.

All setting-out pegs should be outside the excavation area, and should relate to face lines of brickwork or the centre lines of stanchions in steel or reinforced-concrete framed buildings.

In most specifications there is a clause stating that all setting out shall be checked by the Architect or Engineer, but that such checking will not absolve the contractor from responsibility for the accuracy of the work. It is essential, therefore, that all setting out should be carefully and accurately carried out, since mistakes found after construction has started are expensive to put right.

All setting-out pegs should be clearly marked with a 50 mm band of red paint round the top (level pegs should have a band of blue) and information painted on the side stating to what the pegs refer, i.e. stanchion centre line, face of wall, M.H. No., etc. The ganger should know exactly how to work to these pegs before construction starts, as small misunderstandings on his part can lead to serious and costly mistakes.

SETTING OUT BUILDINGS USING TAPES ONLY

In setting out the house shown in fig. 151 first determine the projection of the foundations from the face of the wall, say in this case it is a distance of 270 mm. To keep the pegs undisturbed by the foundation excavation, they should be about 0·6 m from the face of the wall. Although it is possible to set out the right angles by using the 3, 4, 5 rule, as most foremen do, it is better to calculate the diagonal from peg 2 to peg 3; this will speed up the work and lessen the chances of inaccuracy. The method of procedure is as follows: pegs 1 and 2 are fixed on the base line, pegs 3 and 4 by taping from pegs 1 and 2, pegs 5 and 6 are lined in between pegs 1 and 3 and pegs 2 and 4. (The order of placing the pegs is shown by their numbering in fig. 151, i.e. the first peg to be placed is peg number 1, the second peg number 2 and so on). Pegs 3, 4, 5 and 6 are the main setting-out pegs, and profiles are erected round these pegs (see fig. 152). Pegs 1 and 2 are subsidiary pegs which can be removed after the work has been checked.

With the aid of a level and staff a mark may be placed on the main

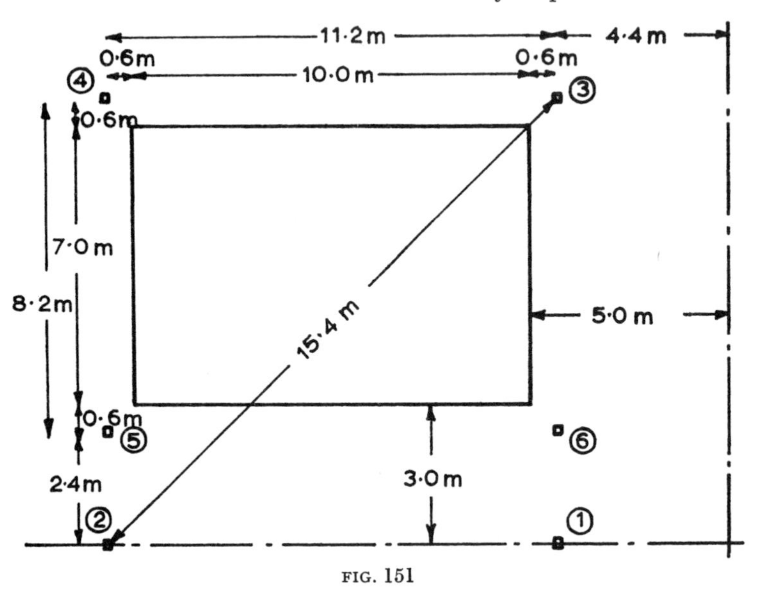

FIG. 151

pegs at floor-level or some multiple of 150 mm above or below the floor-level. The profile boards are then erected level with the mark and used as sight rails.

Hemp lines are tied from profile to profile to give the foundation trench and wall lines (as shown in fig. 152).

When the building is not a rectangle (as in fig. 153), treat the building as a rectangle of the overall dimensions, and erect subsidiary profiles for the return and set-back walls.

SETTING OUT FRAMED BUILDINGS

The erection of steel work is usually the work of a specialist subcontractor, but it is generally the duty of the main contractor to fix the bases and holding-down bolts. Since the position of the bases have to be guaranteed to a tolerance of $+3$ mm, it is essential that accurate

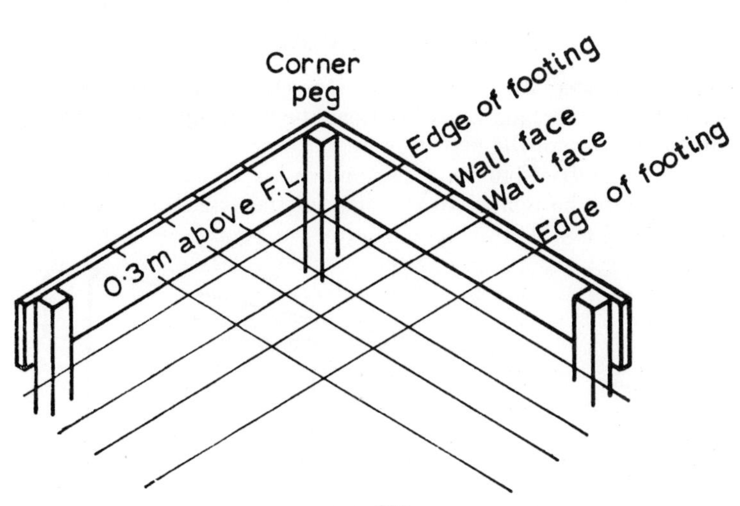

FIG. 152

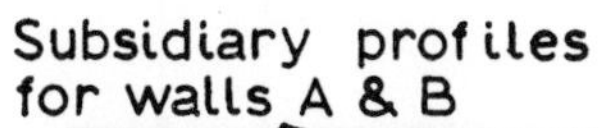

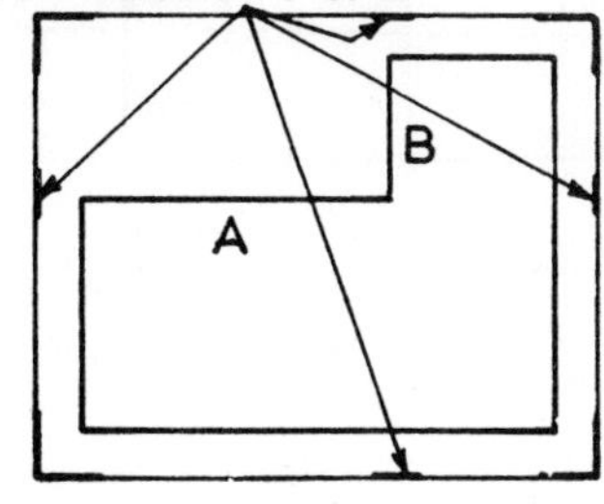

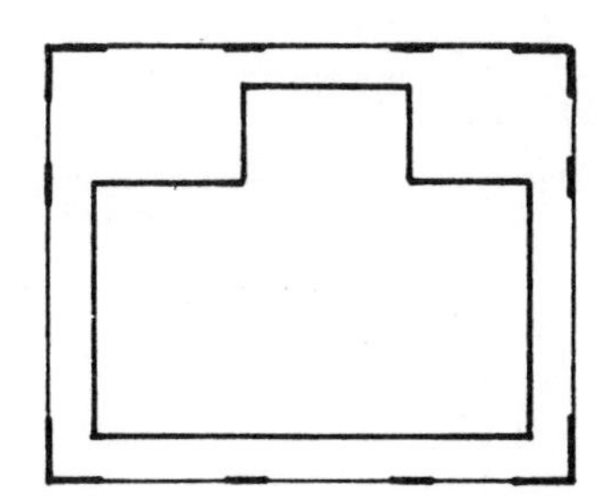

FIG. 153

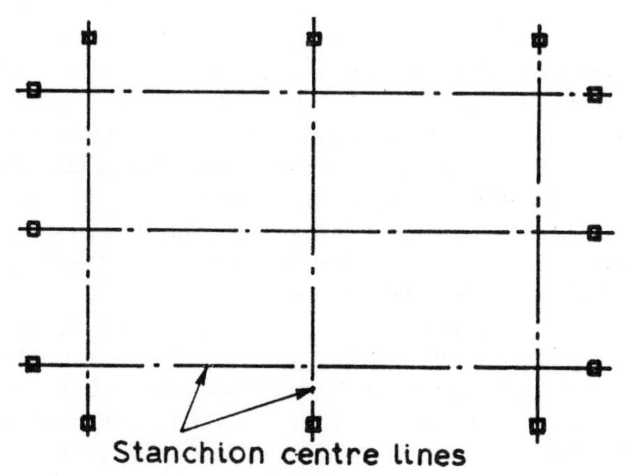

FIG. 154

theodolites and steel tapes be used in the setting out. Stout pegs are inserted on the stanchion centre lines as shown in fig. 154, and a nail knocked in the top of the peg, dead on the centre line.

When an industrial building is set out on a city site, the stanchion centre lines may be related to the original traverse survey lines (see fig. 155).

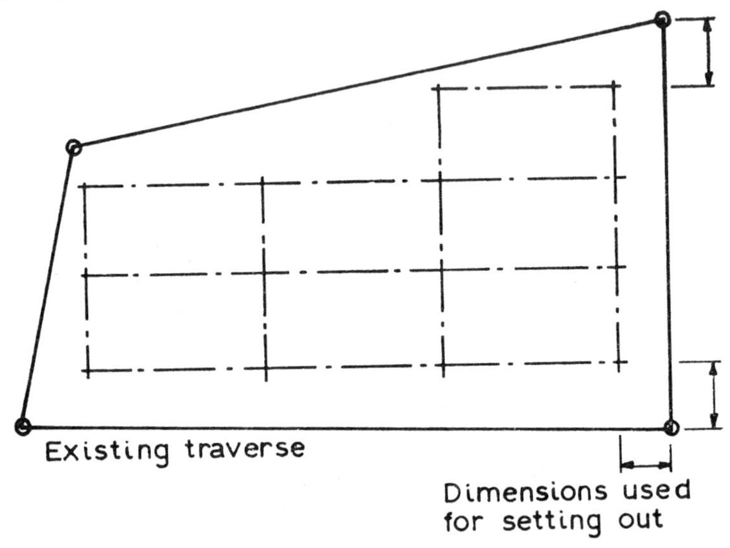

FIG. 155

SETTING OUT BUILDINGS ON LARGE INDUSTRIAL ESTATES

A large site having a number of buildings to be set out should be first gridded. A grid is a series of parallel lines, spaced equidistantly apart, crossed at right angles by another series of parallel lines spaced equidistantly apart. The use of the grid lines will enable the buildings to be located more quickly and accurately than working from one base line for the whole site. A grid, with the lines, say, 100 m apart, is drawn to scale on tracing-paper. The tracing paper is then laid over the site layout on which is marked the position of buildings, services, possible storage areas, ponds and other obstructions. The tracing-paper is moved about so that the intersections of the grid lines (grid stations) do not coincide with the buildings, etc. The grid stations are pricked through on to the layout, and the grid lines drawn on it. The centre lines of the buildings are produced until they intersect the grid lines, and the distances from the intersections to the grid stations scaled or calculated and the distances marked on the layout. The grid is given an arbitrary north–south meridian to coincide with one of the grid lines and the stations co-ordinated (see fig. 157). The bearing of the grid to a long straight road or railway line, or its position relative to a fixed object, is determined from the layout. The grid is then set out on the site, and permanent points established at the grid stations (see fig. 156). To establish a *permanent point* a length of metal tube or bar of about 30 mm diameter is driven in at the required position and level, the earth dug from around it and the hole filled with concrete. The reduced level of the top of the tube and its position are masked on the concrete before it sets. The permanent point can also be used as a site bench mark.

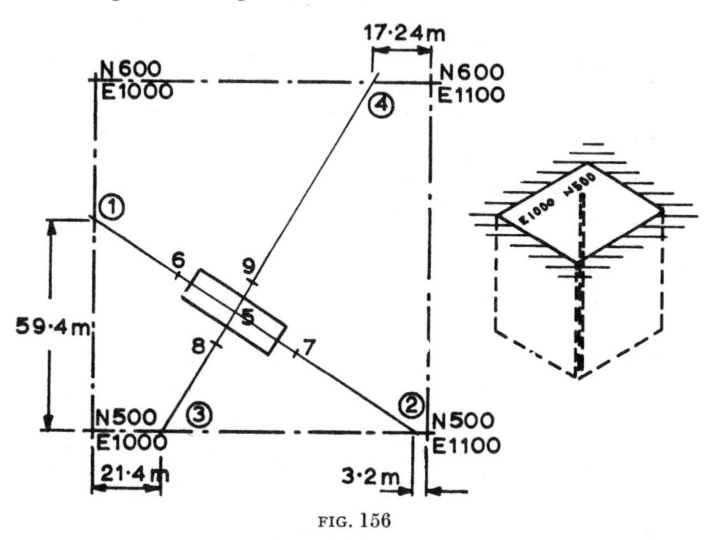

FIG. 156

To set out the building shown in fig. 156, set up the theodolite at N. 500, E. 1000, sight to N. 600, E. 1000, and measure out 59·4 m and insert peg 1. Pegs 3 and 2 are fixed by sighting on to N. 500, E. 1100, peg 4 by setting up at N. 600, E. 1100 and sighting on to N. 600, E. 1000. Set up on peg 4, sight on to peg 3, place a peg on either side of the intersection of lines 1–2 and 3–4. Set up on peg 1, sight on to peg 2 and place peg 5 which intersects line 3–4 and is on the line of 1–2. Set up on peg 5 and place pegs 6, 7, 8 and 9 at some whole number of feet from the stanchion centre line and clear of the excavation.

The grids are also useful for setting out roads and other services, because important points can be quickly located from the layout plan. Points such as T.P.s and I.P.s can be co-ordinated (see fig. 157).

Before going to set out the building, the name or number of the building, its location, details of the centre to centre of the stanchions and floor-level should be noted in a setting-out book. A common form of setting-out book is the same size as a chaining field-book, and the leaves are of graph paper. The location plan is placed on one page, and the setting-out details on the facing page (see fig. 158).

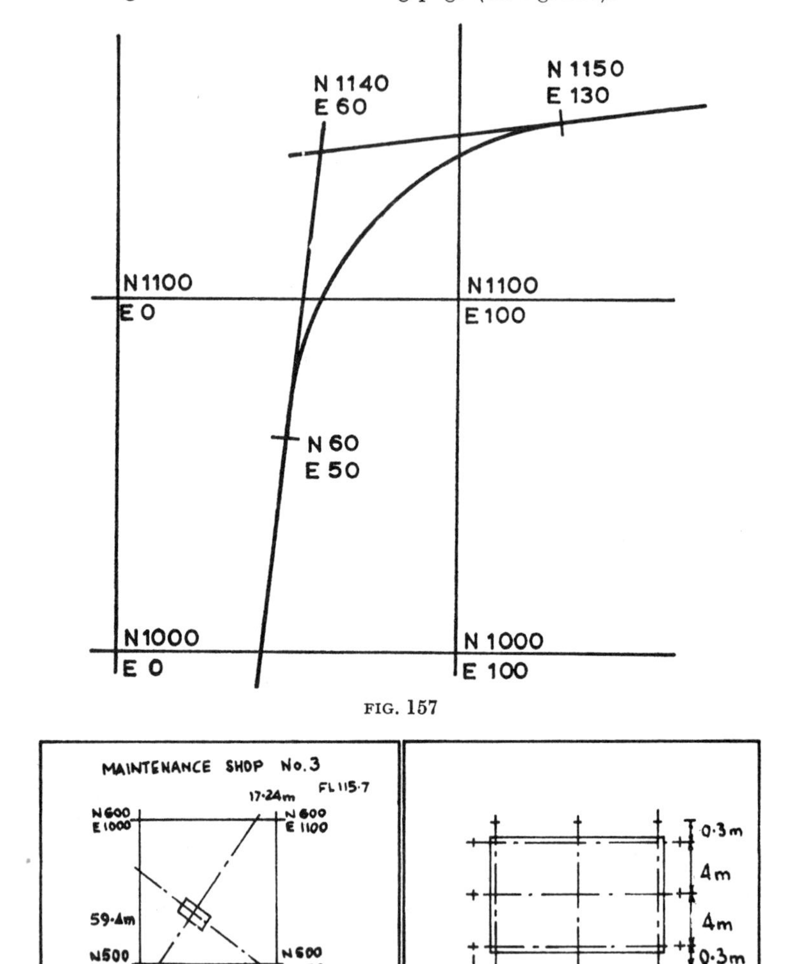

FIG. 157

FIG. 158

MEASURING UP EXISTING BUILDINGS

Surveys of existing buildings are often required for such purposes as alterations, maintenance, valuations, etc. Efforts should be made before surveying to determine whether there are any plans of the building in existence which may be held by the owners or the local authority.

It is important that a system of measurement is adopted, otherwise vital dimensions will be missed, necessitating a further visit to the site. The method set out below has proved to be successful in practice.

1. All sketches should be made on squared paper and roughly to scale.

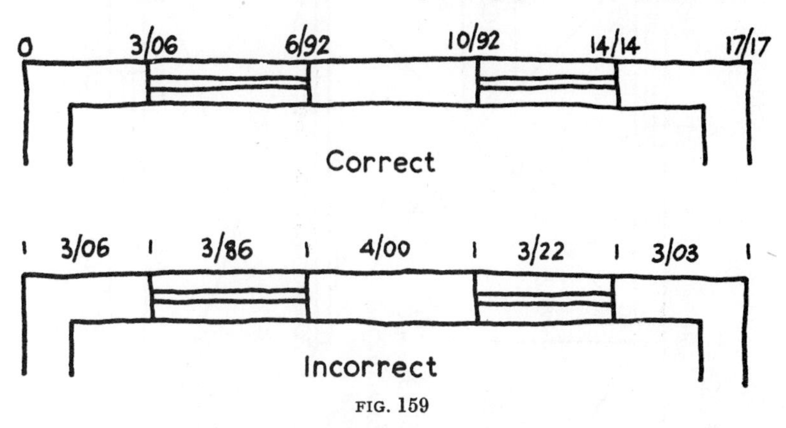

FIG. 159

2. Whenever possible, running dimensions (see fig. 159) should be taken in preference to a string of dimensions, for if one of the strings has been noted incorrectly it will throw all succeeding dimensions out of position.

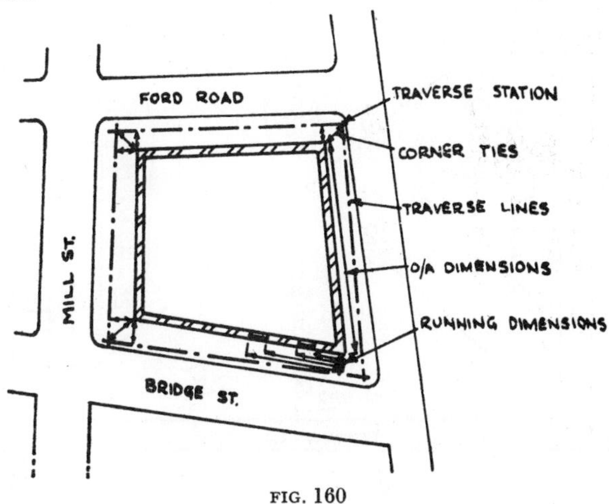

FIG. 160

3. When setting down dimensions, it is better to write them thus 10/60 rather than 10·60, as it is then impossible to confuse the position. of the decimal point.

4. **Procedure**—Sketch an outline plan of the building, together

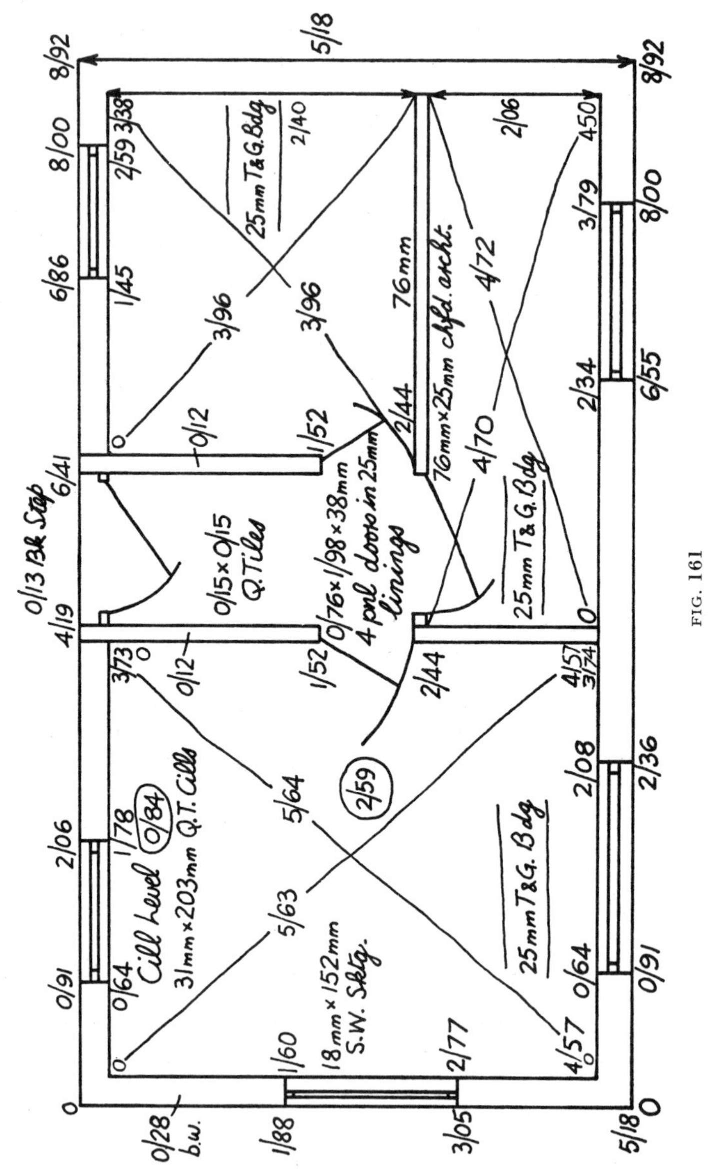

FIG. 161

with any roads, footpaths, etc., enclosing the site, and mark on all relevant information concerning these.

Establish a traverse round the site, tieing in the boundary and the corners of the building (see fig. 160).

Take overall and running dimensions of the external wall at ground-level. The length of the running dimension should equal the overall dimension and the distance, when plotted, of the tied-in corners of the building. In this way the work can be checked.

Determine the thickness of exterior walls by measuring through open doors or windows.

Measure interior starting with the ground floor. Letter the corridors and name or number the rooms. In the case of a small building one sketch will be sufficient for each floor (see fig. **161**), but in the case of a larger structure several sketches may be required. In each room take the following dimensions:

(*a*) Diagonals from corner to corner as a check on the shape of the room. It is sometimes found that rooms which to the eye appear square are in fact not so.

(*b*) Overall dimensions of walls.

(*c*) Running dimensions of walls.

(*d*) Overall heights noted and enclosed in a circle on sketch.

(*e*) Running heights, picking up heights of doors, sills, lintels, etc., noted and enclosed in a circle on the sketch.

Using same procedure, measure up any basements; then move up the building floor by floor to the roof.

In the roof determine the heights of the gables and pitch of the roof.

To determine the thickness of the floors, measure the floor-to-floor distance by measuring from staircase landings or lift shafts.

To obtain a running vertical dimension for the building, measure from the window sill to pavement at each floor.

More detailed plans and elevations are often required of certain details of the construction, and these should be made on the same sheet as the sketch of the room to which they are related.

It is sometimes an advantage to take photographs of parts of the building which are inaccessible for measuring or are of fine, intricate detail, such as carved cornices, etc.

It is advisable, if the survey is some distance from the office, not to leave the site until a rough plot of the survey is made. If this is done, any dimension inadvertently overlooked may be taken, and no time will be wasted in returning to the site to take the dimension later.

Example

It is required to set out on the site the position of a new building. Make sketches showing the method of setting out the pegs and profiles for a rectangular building. Show markings on the boards to be used as guides for setting out:

(*a*) the external walls and their foundations; and

(*b*) the centre lines of the columns.

Solution

The method of setting out pegs and profiles has been discussed on page 160. The markings for walls and foundations are usually V saw cuts in the profile board, and the markings for centre lines of columns are usually made with nails.

INDEX

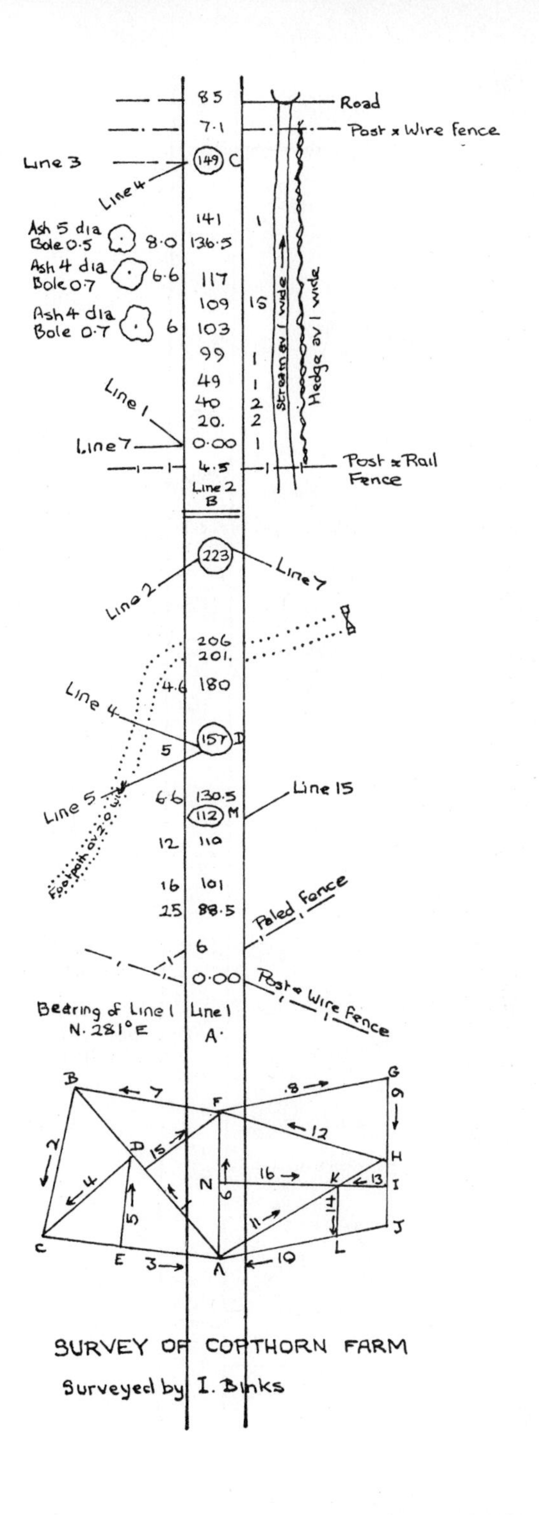
Road
Post & Wire fence
Line 3
Line 4
Ash 5 dia Bole 0.5
Ash 4 dia Bole 0.7
Ash 4 dia Bole 0.7
Stream av 1 wide
Hedge av 1 wide
Line 1
Line 7
Post & Rail Fence
Line 2
B
Line 2
Line 7
Line 4
Line 5
Footpath av 2.0 wide
Line 15
Paled Fence
Post & Wire Fence
Bearing of Line 1 N. 281° E
Line 1
A
SURVEY OF COPTHORN FARM
Surveyed by I. Binks

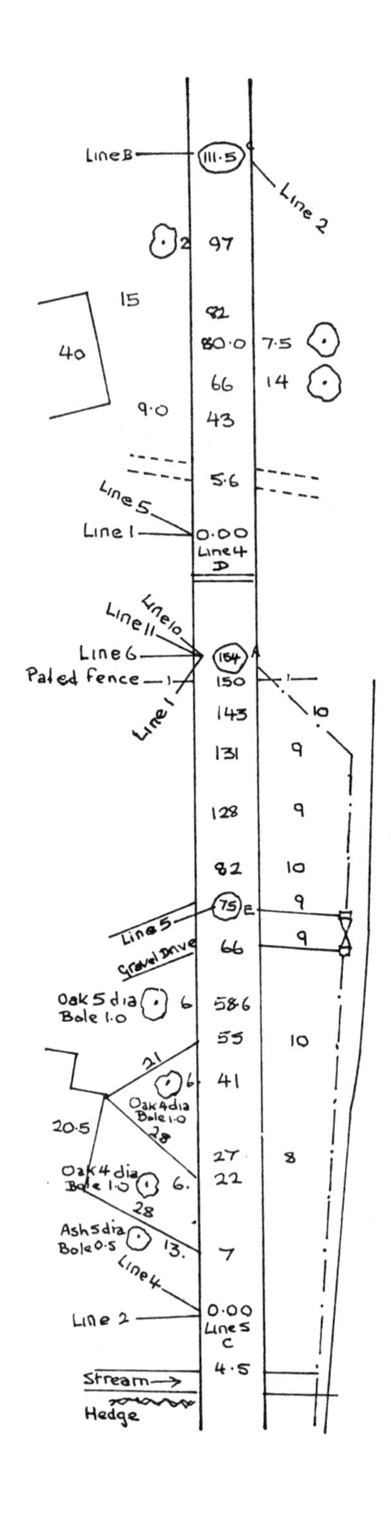

Line B
111·5
Line 2
2
97
15
82
40
80·0
7·5
66
14
9·0
43
5·6
Line 5
Line 1
0·00
Line 4
D
Line 10
Line 11
Line 6
154
A
Paled fence
1
150
1
Line 1
143
10
131
9
128
9
82
10
75
E
9
Line 5
9
Gravel Drive
66
Oak 5 dia
Bole 1.0
6
58·6
55
10
21
6
41
Oak 4 dia
Bole 1.0
28
20·5
27
8
Oak 4 dia
Bole 1.0
6.
22
28
Ash 5 dia
Bole 0·5
13.
7
Line 4
Line 2
0·00
Line 5
C
4·5
Stream
Hedge

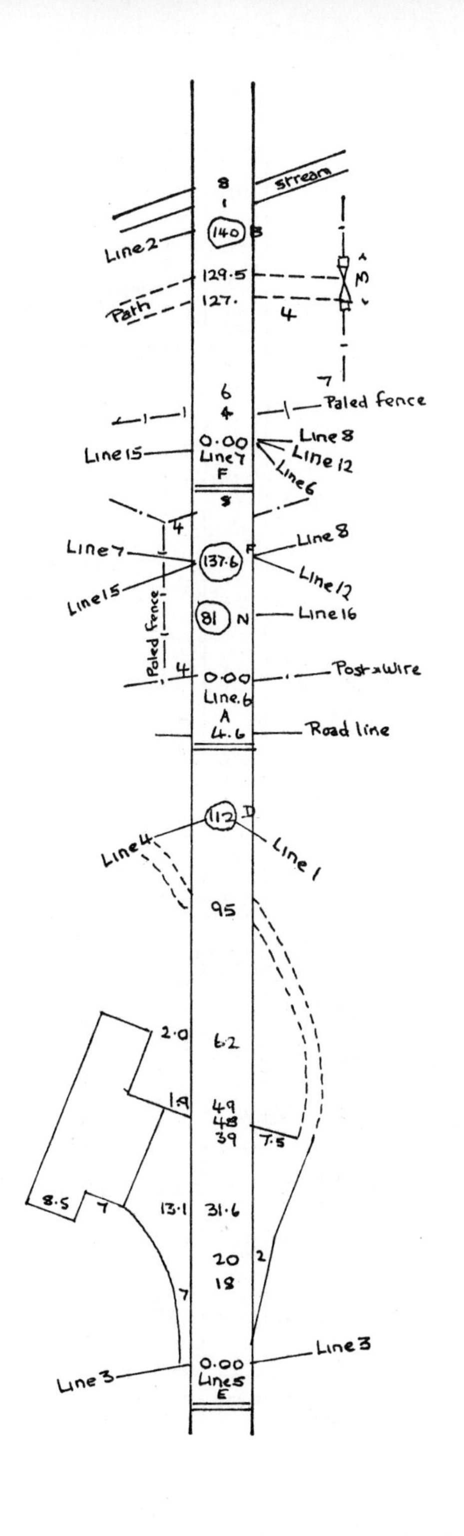
stream
8
1
140
B
Line 2
129.5
127.
Path
4
M
7
6
4
Paled fence
0.00
Line 7
F
Line 15
Line 8
Line 12
Line 6
8
4
Line 7
137.6
F
Line 8
Line 12
Line 15
81
N
Line 16
Paled fence
4
0.00
Line. 6
A
Post & Wire
4.6
Road line
112
D
Line 4
Line 1
95
2.0
6.2
1.9
49
48
39
7.5
8.5
7
13.1
31.6
20
2
18
7
Line 3
0.00
Line 5
E
Line 3

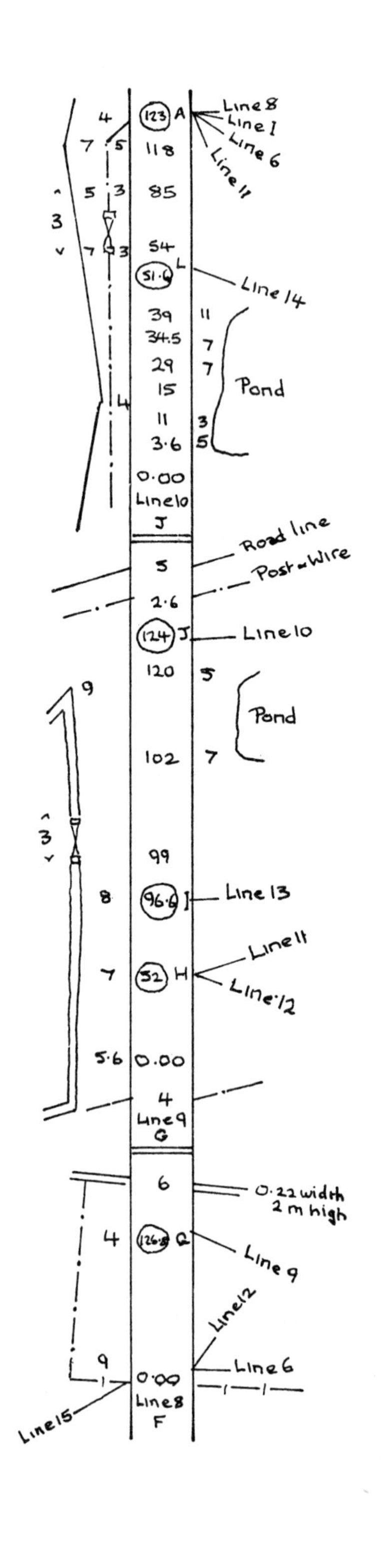

Line 8
Line 1
Line 6
Line 11
123 A
118
85
54
51.6 L
Line 14
39
34.5
29
15
11
3.6
Pond
0.00
Line 10
J
Road line
Post & Wire
5
2.6
124 J
Line 10
120
Pond
102
99
96.6 J
Line 13
Line 11
52 H
Line 12
0.00
4
Line 9
G
6
0.22 width
2 m high
126.8 G
Line 9
Line 12
Line 6
0.00
Line 8
F
Line 15

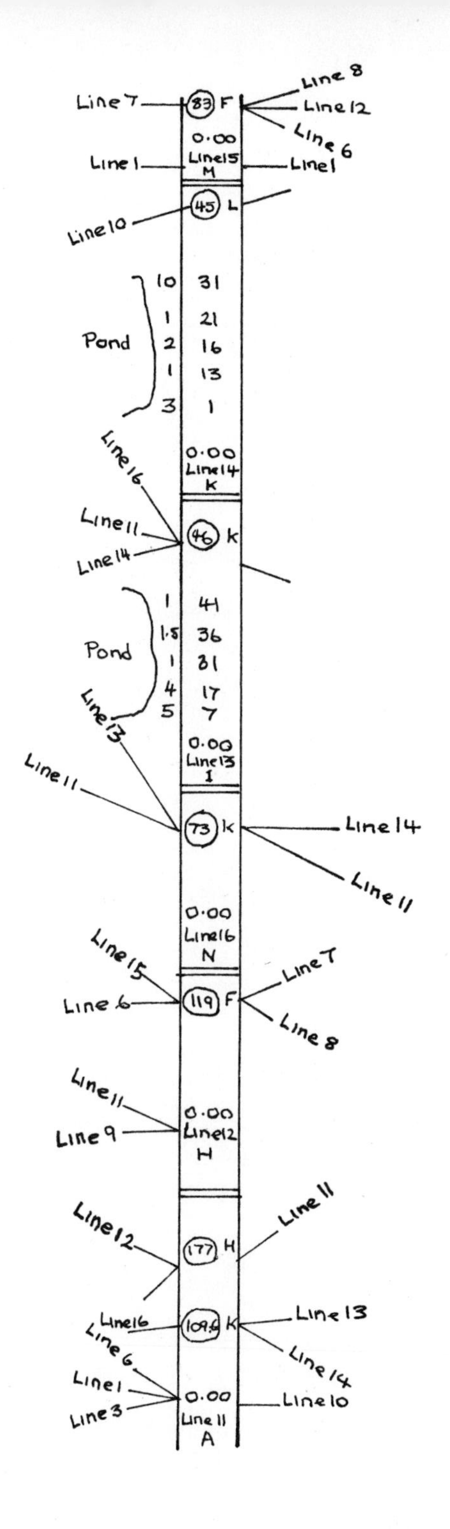

Line 7
83
F
Line 8
Line 12
Line 6
0.00
Line 15
M
Line 1
Line 1
45
L
Line 10
10
31
1
21
Pond
2
16
1
13
3
1
0.00
Line 14
K
Line 16
Line 11
46
K
Line 14
1
41
1.5
36
Pond
1
31
4
17
5
7
Line 13
0.00
Line 13
I
Line 11
73
K
Line 14
Line 11
0.00
Line 16
N
Line 15
Line 6
119
F
Line 7
Line 8
Line 11
0.00
Line 9
Line 12
H
Line 12
177
H
Line 11
Line 16
109.6
K
Line 13
Line 6
Line 14
Line 1
0.00
Line 10
Line 3
Line 11
A